Ecological-Economic Modelling fo Conservation

Both ecologists and economists use models to help develop ~~~~~~ biodiversity management. The practical use of disciplinary models, however, can be limited because ecological models tend not to address the socioeconomic dimension of biodiversity management, whereas economic models tend to neglect the ecological dimension. Given these shortcomings of disciplinary models, ecological and economic knowledge need to be combined into ecological-economic models. Gradually guiding the reader into the field of ecological-economic modelling by introducing mathematical models and their role in general, this book provides an overview of ecological and economic modelling approaches relevant for research in the field of biodiversity conservation. It discusses the advantages and challenges associated with ecological-economic modelling, together with an overview of useful ways of integrating ecological and economic knowledge and models. Although this is a book about mathematical modelling, ecological and economic concepts play an equally important role, making the book accessible for readers from very different disciplinary backgrounds.

MARTIN DRECHSLER is a senior scientist at the Helmholtz Centre for Environmental Research – UFZ, Germany. His research includes the mathematical modelling of populations in fragmented and dynamic landscapes, the mathematical ecological-economic analysis of instruments and strategies for biodiversity conservation, and the model-based assessment of renewable energy deployment as well as mathematical optimisation and decision theory. He originally trained as a physicist, which provided him with the necessary mathematical background to understand, develop and analyse mathematical models. Some 20 years ago, he began collaborating with economists to contribute to the development of the research field of ecological-economic modelling. Today he is one of the most prolific authors in this field.

ECOLOGY, BIODIVERSITY AND CONSERVATION

The world's biological diversity faces unprecedented threats. The urgent challenge facing the concerned biologist is to understand ecological processes well enough to maintain their functioning in the face of the pressures resulting from human population growth. Those concerned with the conservation of biodiversity and with restoration also need to be acquainted with the political, social, historical, economic, and legal frameworks within which ecological and conservation practice must be developed. The new Ecology, Biodiversity, and Conservation series will present balanced, comprehensive, up-to-date and critical reviews of selected topics within the sciences of ecology and conservation biology, both botanical and zoological, and both 'pure' and 'applied'. It is aimed at advanced final-year undergraduates, graduate students, researchers, and university teachers, as well as ecologists and conservationists in industry, government, and the voluntary sectors. The series encompasses a wide range of approaches and scales (spatial, temporal and taxonomic), including quantitative, theoretical, population, community, ecosystem, landscape, historical, experimental, behavioural, and evolutionary studies. The emphasis is on science related to the real world of plants and animals rather than on purely theoretical abstractions and mathematical models. Books in this series will, wherever possible, consider issues from a broad perspective. Some books will challenge existing paradigms and present new ecological concepts, empirical or theoretical models and testable hypotheses. Other books will explore new approaches and present syntheses on topics of ecological importance.

Ecology and Control of Introduced Plants
Judith H. Myers and Dawn Bazely

Invertebrate Conservation and Agricultural Ecosystems
T. R. New

Ecological-Economic Modelling for Biodiversity Conservation

MARTIN DRECHSLER

Helmholtz Centre for Environmental Research – UFZ

CAMBRIDGE
UNIVERSITY PRESS

CAMBRIDGE
UNIVERSITY PRESS

University Printing House, Cambridge CB2 8BS, United Kingdom

One Liberty Plaza, 20th Floor, New York, NY 10006, USA

477 Williamstown Road, Port Melbourne, VIC 3207, Australia

314–321, 3rd Floor, Plot 3, Splendor Forum, Jasola District Centre, New Delhi – 110025, India

79 Anson Road, #06-04/06, Singapore 079906

Cambridge University Press is part of the University of Cambridge.

It furthers the University's mission by disseminating knowledge in the pursuit of education, learning, and research at the highest international levels of excellence.

www.cambridge.org
Information on this title: www.cambridge.org/9781108493765
DOI: 10.1017/9781108662963

First published 2020

Printed in the United Kingdom by TJ International Ltd, Padstow Cornwall

A catalogue record for this publication is available from the British Library.

Library of Congress Cataloging-in-Publication Data
Names: Drechsler, Martin, 1966- author.
Title: Ecological-economic modelling for biodiversity conservation / Martin Drechsler.
Description: Cambridge ; New York, NY : Cambridge University Press, 2020. | Series: Ecology,
    biodiversity and conservation | Includes bibliographical references and index.
Identifiers: LCCN 2019038235 (print) | LCCN 2019038236 (ebook) | ISBN 9781108493765
    (hardback) | ISBN 9781108725514 (paperback) | ISBN 9781108662963 (epub)
Subjects: LCSH: Biodiversity conservation–Mathematical models. | Biodiversity conservation–
    Econometric models. | Ecosystem management–Statistical methods. | Environmental economics.
Classification: LCC QH77.3.S73 D74 2020 (print) | LCC QH77.3.S73 (ebook) |
    DDC 333.95/16–dc23
LC record available at https://lccn.loc.gov/2019038235
LC ebook record available at https://lccn.loc.gov/2019038236

ISBN 978-1-108-49376-5 Hardback
ISBN 978-1-108-72551-4 Paperback

Additional resources for this publication at www.cambridge.org/drechsler.

Contents

Preface

Despite various efforts to halt or reverse the current trend, biodiversity is being lost at an alarming rate across the world. At first sight, biodiversity may be regarded as an ecological issue and a topic of ecological research. However, its loss has economic causes and economic consequences, and economists are increasingly interested in the economic dimension of the loss and the conservation of biodiversity. To encompass the full complexity of biodiversity, both its ecological and economic dimensions must be considered in an integrated manner, ideally even together with other scientific disciplines such as hydrology, climatology, sociology, psychology and philosophy. Among various concepts for interdisciplinary integration, ecological-economic modelling has proven very fruitful and is gaining relevance and popularity both among ecologists and economists.

The present book provides an overview of the state of the art of ecological-economic modelling. The focus here is on mechanistic process models that model the relationships between causes and consequences through mathematical rules or equations. Statistical models such as habitat suitability and species distribution models that explain species presence from biotic and abiotic conditions, or econometric models that for instance explain human behaviour through environmental and socioeconomic variables, are not covered in this book. This is not because statistical models cannot be used for ecological-economic modelling – in fact there are a number of ecological-economic models that contain statistical models as components – but, in the author's view, to date mechanistic process models form the majority of ecological-economic models, and a fair consideration of statistical models would be beyond both the author's knowledge and the scope of this book.

To build integrated ecological-economic models, both conceptual knowledge (about what to integrate and for what purpose) and formal mathematical methods (how to integrate the available knowledge) are required. This book tries to address both dimensions of the integration process by considering both conpceptual thinking and mathematics on intermediate levels. The book thus contains more mathematics and less conceptual thinking than standard books on ecology or on environmental or ecological economics, but it contains less mathematics and more conceptual thinking than standard books on mathematical and complex-systems modelling. In this way, the book is integrative not only with regard to the disciplines of ecology and economics but also with regard to the mediation between concepts and their formal mathematical implementation. A particular feature of the book is the employment of numerous modelling examples from the literature whose selection is, of course, subjective but carried out in an attempt to be instructive and to cover a wide range of concepts and methods.

Although generally not stated explicitly, the focus of the book – especially the literature examples – is on terrestrial biodiversity conservation. Some of the modelling approaches addressed here can also be applied to marine and freshwater ecosystems, but the inclusion of these ecosystems would be beyond the book's scope.

Furthermore, in most of the conservation problems considered biodiversity has no market value, and the conservation of biodiversity requires particular policies and strategies that differ from those applicable to marketable natural resources such as timber and fish.

Since biodiversity often has no market value, the economic valuation of biodiversity (and the environment in general) is a major research field in environmental and ecological economics, covered in various books. In economic terms, valuation deals with society's demand for biodiversity. In contrast, much of the literature on ecological-economic modelling deals with the supply of biodiversity and, in particular, the question of how limited financial resources should be spent cost-effectively to maximise biodiversity levels for a given cost or budget. Most of the present book focuses on that supply side of biodiversity, although a few sections address the demand side, as well. It will be argued that the combined consideration of both sides is very fruitful.

Before proceeding to an outline of the structure of the book, some remarks should be added concerning the book's intended readership. Because this is book about mathematical modelling, some mathematical knowledge is required to understand the models presented. However, it is not necessary to understand all the equations in detail to capture their meaning and the modelling concepts behind them. The primary audience of the book is researchers and graduate students who already have a proficient knowledge base in mathematical modelling, quantitative ecology and economics, or at least one of these three disciplines, and who wish to broaden their knowledge beyond their own discipline in order to work with researchers from other disciplines, or even develop integrated models on their own. This is not a textbook in the narrow sense but, rather, a compendium of relevant ecological-economic models, concepts and approaches. However, it can also be used as a textbook for graduate courses, and in fact some of its content is based on a lecture about ecological-economic modelling held by the author regularly over the past few years.

The book is organised into four parts. Part I introduces mathematical modelling in general, describing in particular the various purposes models can have, as well as typical model features. Modelling examples are considered from the disciplines of physics, ecology and economics.

Part II provides an overview of ecological models relevant for biodiversity conservation. After an introductory chapter, three important model features are addressed: stochasticity, spatial structure and individual variability and behaviour. In a final chapter the modelling approaches of the previous chapters are combined to discuss one of the central questions of biodiversity research: why and how can different species coexist?

The economic side of ecological-economic modelling is presented in Part III, which starts with basic concepts of environmental economics, such as biodiversity loss as a market failure and policy instruments for mitigating that market failure. As a fundamental approach for policy analysis, the following chapter presents basics of game theory, which is followed by a chapter on incentive design and a chapter on the modelling of human behaviour and decisions. The final chapter applies concepts derived in the previous chapters to discuss recent research on the agglomeration bonus – a policy instrument for incentivising spatially coordinated conservation efforts by landowners.

The final section, Part IV on ecological-economic modelling, starts with a brief summary of the history of economic thought, including the foundations of environmental economics and ecological economics, to derive recommendations for the design of ecological-economic models. The following chapter deals with the advantages of ecological-economic modelling compared with disciplinary research, as well as associated difficulties and challenges. After a chapter about major approaches for the integration of ecological and economic models, two examples are presented in which ecological-economic models are used to analyse policy instruments for the conservation of species. Part IV concludes with an outlook on the possible future of ecological-economic modelling.

Acknowledgements

Many people have positively influenced my research and thereby contributed to this book. In addition to the many colleagues in the Department of Ecological Modelling at the UFZ and various national and international research projects, I want to highlight a few of them by name. The first is Christian Wissel, supervisor of my first PhD thesis and my former boss. I am most grateful for the amount of time he has invested in discussions with me about my research and for his support when I later took my first steps in the new field of ecological-economic modelling. My luck with excellent supervisors continued when I was a postdoc with Mark Burgman at the University of Melbourne. In Mark's lab I learned a lot about the role of modelling and quantitative methods in ecology and developed confidence in my own abilities as a modeller and theoretical ecologist. Third is my long-standing colleague and good friend Frank Wätzold who introduced me to economics and with whom I have developed many ecological-economic models. Many of the research questions behind these models were raised by Frank. I very much look forward to more joint research projects with him in the future. I am particularly indebted to Frank for carefully reading an earlier version of this manuscript and providing helpful comments, and for the same help I want to thank my colleague Volker Grimm who has also been a good advisor in many circumstances and has critically commented on several of my papers, starting from the very first ones. I have become so accustomed to a friendly and cooperative working atmosphere that I almost forget that such a thing cannot be taken for granted; to a great extent this is the achievement of my boss, Karin Frank. Lastly I want to thank my wife, Gabriele Körner, who may not have contributed significantly to my scientific life but certainly to my private life, and undoubtedly there is beneficial feedback between these two.

Part I
Modelling

1 · *What Is a Model?*

If one put this question to travellers walking through Leipzig Main Station it's likely that quite a few of them would mention the names of some popular fashion models. But some others would probably point to a glass case in the middle of the hall which contains a small landscape with plastic houses, cars, people and trains running around in circuits.

While installed to entertain people, the glass case may help to give us a first answer to the question 'what is a model?' The items in the case are obviously meant to represent a socio-technical system that can be found in the real world. The model railway is smaller than the real-world system and also lacks various details, such as the smoke that would come out of a real steam engine, and the human agents in the landscape seem to be frozen in their activities. Thus, the model railway abstracts from a number of details present in the real world. However, it is easy for the spectator to imagine the smoke of the engines and understand what the human agents would be doing if they could move. So the level of abstraction is obviously well chosen: if there were much more detail and less abstraction, the construction of the model would have been too costly and the model too difficult to run; while if there were much more abstraction, running and watching the model would probably be boring and the model would not fulfil its purpose of entertaining people.

The model railway includes several of the issues and features that characterise a model after Baumgärtner et al. (2008, p. 389): namely, that a model is designed to serve a certain purpose and that it is an abstract representation of a real system. To capture the remaining characteristics of such a model one might leave Leipzig Main Station and walk to the nearby tram stop. Several maps are installed there to guide locals and tourists. One of these maps looks similar to Fig. 1.1, showing streets, buildings and other features of Leipzig's city centre.

Like the model railway, the map contains an appropriate level of abstraction to fulfil a given purpose: transmitting information about the spatial structure of Leipzig's city centre from the producer of the map to

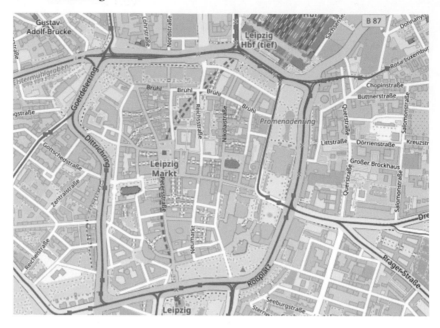

Figure 1.1 Map of the city centre of Leipzig. Source: OpenStreetMap (URL: www
.openstreetmap.org/#map=15/51.3418/12.3787; last access 4 February 2019)

its users. In order to approach Baumgärtner et al.'s final model characteristic, one may note that the map in Fig. 1.1 is probably readable only for a
person who has been in a city before. Only a person with this experience
will understand that the linear structures on the map are streets while the
closed shapes depicted in grey represent buildings. In contrast, for someone who has lived their entire life in a remote place without any contact
with the Western world, this will be much less obvious. This demonstrates that a model (in the present case a map) can only be understood if
the user of the model has an understanding of the construction of the
world (in the present case cities) similar to that of the developer of the
model. This brings us to the final characteristic of a model: that it 'is based
on the concepts within a ... community's basic construction of the
world' (Baumgärtner et al. 2008, p. 389).

While most people on earth share the concepts that are required to
read a map, concepts may strongly differ among different communities
when scientific models are considered. Ecological models are often based
on the ecological concept of a population, which in turn is based on the
concepts of individuals and species, since a population by definition can
be formed only by individuals of the same species. In economics, the

concept of a market that mediates demand and supply for goods and services is a central one.

Concepts further differ among the disciplines with regard to the quantities characterising the state of a system. Physics and economics often employ the concept of equilibrium. A system is in equilibrium when all forces or processes cancel each other and there is no change in any component of the system. In economics, such an equilibrium may be reached in a market when demand and supply for a good are equal, so that the quantity of the good and its price are constant. For a certain time, the concept of equilibrium was popular in ecology, too. However, it soon came to be regarded as insufficient, since ecological systems are usually not in equilibrium. Instead, stability concepts such as persistence (how long does a system persist within a certain state?) and resilience (how difficult is it to push the system away from a certain state to another state through some external driving force, and how long does it take for the system to resume its original state after the driving force has been turned off?) have been developed and are more appropriate to characterise the state and dynamics of an ecological system (Grimm and Wissel 1997). Over the past 100 years or so, these other stability concepts have become accepted in physics, too, and more recently also in economics (Perrings 2006).

Altogether, the concepts within a scientific community's basic construction of the world differ between disciplines and partly overlap. When building interdisciplinary models such as ecological–economic models, it is necessary to take this issue into account carefully. To summarise and conclude this chapter, I now combine the characteristics of a model that have already been introduced and quote Baumgärtner et al. (2008, p. 389):

A model is an abstract representation of a system under study, explicitly constructed for a certain purpose, and based on the concepts within a scientific community's basic construction of the world that are considered relevant for the purpose.

2 · *Purposes of Modelling*

Chapter 1 emphasised the important role of purpose in the adequate design of a model. It concluded that the clear formulation of purpose is an important step before and during the development of a model.

Model purposes are manifold. Baumgärtner et al. (2008) distinguish nine different purposes: theory development, generalisation, theory testing, understanding, explanation, prediction, decision support, communication and teaching. I will add two further purposes: integration of knowledge and mediation between scales. Lastly, models may be distinguished by being general or specific and by being used in positive or in normative analysis. I will introduce each of these purposes using examples from physics, ecology and economics.

2.1 Theory Development

The following three sections address the relationship between models and theories. The topics of these three sections – theory development, generalisation and theory testing – are interrelated. In the development of a theory it is important to consider that the theory needs to be able to generalise specific observations into a coherent framework and that it can be tested against real-world observations, because only then will it be of value.

Models can form an intermediate element between an abstract theory and the observable world, or as Morrison and Morgan (1999a) put it, models act as mediators between theory and the world. This is possible because they contain elements from both sides (Morrison and Morgan 1999b). Using a term from computer science, I would add that a model can serve as an instance, that is, a realisation, of the theory in the observable world.

To explain these definitions of a model I will use a theory from the realm of physics. Before discussing the theory, however, I want to illustrate it by reference to a social phenomenon that can be observed in some cities: that people form distinct neighbourhoods or groups with strong similarities within groups but strong dissimilarities between groups. The first and probably most famous model analysis that addressed this issue was by Schelling (1969), who wondered why many cities in the United States consist of distinct black and white neighbourhoods. Schelling hypothesised that this could be due to people's preferences to be surrounded by their like, and that people who are currently in the minority within their neighbourhood would move to another neighbourhood in which their colour is the majority. If we code the colours black and white into two numbers, such as +1 and −1, and assume that a person's 'happiness' increases by an amount g for each neighbour of the same colour and decreases by g for each neighbour of a different colour, we can write the happiness h of citizen i with colour s_i ($s_i \in \{-1, +1\}$) as

$$h_i(s_i) = g \sum_{j \in J_i} s_i s_j, \tag{2.1}$$

where index j applies to everyone j in the neighbourhood J_i. Each match ($s_i = s_j = +1$ or $s_i = s_j = -1$) increases $h_i(s_i)$ by an amount g and each mismatch ($s_i \neq s_j$) reduces it by g. The 'total happiness' in society may thus be calculated as the sum of the individual happiness of all:

$$H = \sum_i h_i(s_i). \tag{2.2}$$

Eqs (2.1) and (2.2) are in fact much older than Schelling's work and were first formulated by the physicist Ernst Ising (1900–98) to model the phenomenon of ferromagnetism. Iron and various other metals are so-called ferromagnets characterised by their ability to assume two different phases or states, a magnetic and a non-magnetic phase. The two phases are separated by a critical temperature T_c, so that at temperatures T below T_c the ferromagnet is magnetic and above it is not. The Ising model represents the ferromagnet by a (usually square) grid in D dimensions where on each grid point i a spin s_i is situated (Chaikin and Lubensky 1995; Hohenberg and Krekhov 2015). This spin can be imagined as a small elementary magnet (like the needle of a tiny compass). While in a real ferromagnet a spin can assume any direction, a spin in the Ising model can, for simplicity, assume only one of two possible directions: up or down, or mathematically more conveniently: $s_i = +1$ or $s_i = -1$.

In a ferromagnet, neighbouring spins try to point in the same direction, so neighbouring s_i tend to be either both +1 or both −1. To model this, Ising formulated the so-called Hamiltonian of the system which represents the system's energy, as

$$H = -g \sum_{<i,j>} s_i s_j, \tag{2.3}$$

where the sum runs over all pairs of neighbouring spins (in a two-dimensional grid there would be four neighbours [north, south, east, west], while in a three-dimensional grid each spin would have six neighbours). Parameter g is a positive constant measuring the strength of the interaction between neighbouring spins. Due to the positivity of g, quantity H is smaller when two neighbouring spins have the same sign (both +1 or both −1) than when they have opposite signs (one +1 and the other one −1).

Using the mathematical technique of functional integration, and ignoring fluctuations (see, e.g., Chaikin and Lubensky 1995; Hohenberg and Krekhov 2015), the so-called free energy of the system with volume Ω can be deduced to

$$F = \Omega \left[a\Psi^2 + b\Psi^4 \right] \tag{2.4}$$

with parameter

$$a = \frac{D}{T}(T - 2gD), \tag{2.5}$$

where D is the number of dimensions of the grid and T is temperature (the meaning of parameter $b > 0$ is irrelevant in the present context). Quantity Ψ is the level of magnetisation in the system (Negele and Orland 1988). The free energy is defined as the energy that can be converted into work at a constant temperature and volume, and physical systems attempt to assume a state that minimises their free energy. For $a > 0$ (cf. Fig. 2.1, solid line) the free energy after Eq. (2.4) has only a single minimum located at $\Psi = 0$ (zero magnetisation), while for $a < 0$ there are two minima with non-zero magnetisation, $\Psi \neq 0$. The value $a = 0$ is the critical point that separates the magnetic phase, $\Psi \neq 0$, from the non-magnetic one, $\Psi = 0$.

Setting a of Eq. (2.5) to zero and solving for T yields the critical temperature,

$$T_c = 2gD, \tag{2.6}$$

so that for low temperatures $T < T_c$ we have $a < 0$ and the system is magnetic, while for $T > T_c$ we have $a > 0$ and the system is non-magnetic. Quantity T_c is the critical temperature at which the

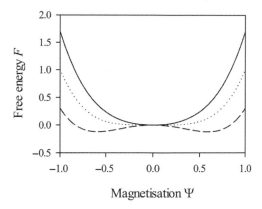

Figure 2.1 Free energy F as a function of the magnetisation Ψ, as given in Eq. (2.4). The parameters are $b = 1$, and $a = 0.7$ (solid line), $a = 0$ (dotted line) and $a = -0.7$ (dashed line).

ferromagnet changes between the magnetic and the non-magnetic phases, a process termed phase transition. As Eq. (2.6) shows, T_c is positively related to the interaction strength g, so that increasing g increases T_c.

The approach of writing the free energy F of a system as a polynomial of some macroscopic system property such as the magnetisation Ψ, solving for the value Ψ^* that minimises F and discussing under which circumstances Ψ^* is non-zero, is the core of the so-called Ginzburg–Landau Theory (GL Theory), named after the two Russian physicists Vitaly Lazarevich Ginzburg and Lev Davidovich Landau. The Ginzburg–Landau Theory provides an intuitive and mathematically relatively simple description of phase transitions in complex systems with many interacting particles or agents (Chaikin and Lubensky 1995). Mathematical calculations allow that theory to be derived from the Ising model, so the model may be regarded as a realisation of the theory. In Section 2.2 we will see that the Ginzburg–Landau Theory has applications in many different types of complex systems.

2.2 Generalisation

In Section 2.1 the Ginzburg–Landau Theory was presented as a suitable tool to describe the phase transition in a ferromagnet. However, the same theory can be used to describe many other phase transitions in very different physical systems, such as the transition from vapour to liquid or vice versa or superconductivity where electrical currents flow without any resistivity if the temperature is below a critical level (Chaikin and Lubensky 1995; Hohenberg and Krekhov 2015). All these systems share the feature that below the critical temperature a macroscopic system property (magnetism in the case of the ferromagnet, the difference between the liquid and gas densities in a vapour–liquid system, and the density of the supercurrent component in a superconductor) is positive.

Phase transitions can also be observed in biological and social systems. The Schelling model is a prominent example. By comparing Eqs (2.1) and (2.2) with Eq. (2.3) it becomes obvious that the 'happiness' H of the people in the Schelling model is (except for the minus sign) formally identical to the Hamiltonian H of the ferromagnet. And in fact, if randomness is added to the people's behaviour in the Schelling model and identified with a temperature (which is quite obvious, considering that temperature results from the random motion of physical particles), the Schelling model exhibits a behaviour very similar to that of the Ising model of the ferromagnet (Stauffer 2007).

These examples demonstrate that the behaviour of very different systems can be described by the same theory and structurally similar models. Although the modelled systems differ (as do the models used for their description: e.g. the Hamiltonian used to model a superconductor differs considerably from the Ising model, Eq. (2.3), above), they can all be cast into the same Ginzburg–Landau Theory. Similarly, in reverse, one may also say that the different models can be regarded as instances of the GL Theory in the observable world.

The Ising model is an example of a model that was originally developed for a physical system but has been subsequently applied to social systems. Another example of such a transfer between disciplines can be found in the analysis of forest fires. The dynamics of forest fires has been a matter of research by both physicists and ecologists (Zinck and Grimm 2009). 'Physical' wildfire models belong to the research fields of statistical physics that deal with the dynamics of complex systems with many interacting particles or agents. These 'physical' wildfire models are usually based on strongly simplifying assumptions about the ecological processes affecting the spatio-temporal dynamics of wildfires and thus have largely been ignored in the ecological literature. 'Ecological' wildfire models, in contrast, consider the ecological processes in much greater detail, which allegedly leads to more realistic model behaviour and better predictions of the fire dynamics in real-world forests. However, the ecological wildfire models are usually too complex and too specific to generate a general understanding of wildfire dynamics or to reveal the general principles governing them.

Zinck and Grimm (2009) compared the dynamics produced by two 'ecological' wildfire and two 'physical' models. All models are grid-based, where each grid cell may be empty or contain one or more trees in a particular condition (more details on grid-based models can be found in Section 6.2). Furthermore, in all models a wildfire can spread through the forest, because a burning grid cell (of course, it's the trees in the cell that burn) can ignite a neighbouring grid cell. In the two 'ecological' models the probability of a grid cell being ignited by a burning neighbour depends on the time since the last fire. This is plausible, because the longer the time since the last fire, the more biomass will have accumulated and the more easily the trees in the grid cell will burn.

The two 'physical' models do not include this burning condition. Instead, in one of these models a grid cell is ignited with certainty if one of its neighbours is burning and if it contains a tree. The second model is a so-called dynamic percolation model (for more details on percolation models, see Section 2.4) in which, as in the first physical wildfire model, grid cells can be ignited by neighbouring grid cells, but where, in contrast, any cell can be ignited – be it occupied by a tree or not.

The authors evaluated and compared the four models with regard to several landscape-scale variables: (i) the relationship between the size A of the burnt area and the frequency $f(A)$ of observing such a fire size, (ii) several parameters (e.g. edge) characterising the geometric shape of a burned area and (iii) the landscape diversity measured by the Shannon–Wiener index (cf. Begon et al. 1990, ch. 17.2.1) of the successional states (which are characterised by the time since the last fire) in the model landscape. It turns out that all models except for the percolation model show the same frequency distribution of burnt areas, which follows a power law:

$$f \sim A^{-\gamma}, \tag{2.7}$$

meaning that larger fires are less frequent than smaller ones. The exponent γ lies between 1.16 and 1.17 for all three models and only the percolation model substantially deviates, with $\gamma = 0.78$. The shape parameters of the burnt areas agree extraordinarily well among all four models and in addition nearly perfectly agree with empirical observations from a forest in Canada. Lastly, the first three models produce nearly the same landscape diversity for many different model parameterisations. Again, only the percolation model deviates slightly from the others.

The reason for the surprising agreement between 'physical' and 'ecological' models is, as the authors argue, that in all of them (except for the percolation model) the probability of a grid cell being ignited by a neighbouring cell depends on the time since the last fire in the cell. While in the two ecological models this feature was included explicitly, the authors show that in the first of the two physical models it emerges from the coupled dynamics of wildfires and regrowth of the trees. Only the percolation model by definition does not include this feature, and the authors emphasise that they included it in the analysis only as some 'kind of null model' to test the validity of their conclusion.

Altogether, the paper by Zinck and Grimm (2009) demonstrates that very different wildfire models can produce similar spatio-temporal dynamics, as long as they share an important feature either explicitly or through emergence: that the probability of a forest stand becoming ignited by a neighbouring fire depends on the stand's age, that is, the time since the last fire in the stand. This feature is likely to be a very general one, and although calling it a theory might be unwarranted today, it could certainly be an essential element within a 'Theory of Wildfires' yet to be developed.

2.3 Theory Testing

As described in Section 2.2, very different wildfire models can produce similar forest fire dynamics, and the generated forest fire patterns agree well with field observations – if the models share an important general feature. Although various general principles have been detected in ecology, it is not easy to find an ecological theory that provides a broad description and understanding of a large class of ecological systems. Grimm (1999) therefore suggests that ecology should use a definition of theory that is different from, for example, physics. While in physics a theory is able to give general answers for a large class of systems (such as the Ginzburg–Landau Theory for phase transitions), theories in ecology cannot provide such general answers, since ecological systems are too different, and so ecological theories should be regarded as frameworks for asking the 'right' questions to analyse ecological systems.

Given the lack of widely applicable ecological theories (and a similar problem with regard to economics, at least as far as it relates to the context of environment and biodiversity), I return to the realm of physics and briefly outline how the Ginzburg–Landau Theory for superconductivity can be verified by way of experimental measurement.

To outline the phenomenon of superconductivity, at sufficiently low temperatures the electrons in a superconductor form pairs which below some critical temperature T_c condense into a so-called superfluid. In this phase the electron pairs no longer interact ('collide') with the atoms of the superconductor and thus can flow without any electrical resistivity.

Since heat tends to break up these electron pairs, one can expect that superconductors with *strongly coupled electron pairs* can sustain superconductivity at higher temperatures (i.e. have a higher T_c) than superconductors with less strongly coupled electron pairs – very similar to Eq. (2.6) where T_c increases with increasing interaction strength g. Indeed, in the more recently detected ceramic 'high-Tc' superconductors the electrons are more strongly coupled than in the metallic 'classical' superconductors with lower T_c. Analysing a superconductor model within the framework of GL Theory, Drechsler and Zwerger (1992) confirmed a positive relationship between T_c and the electron coupling strength.

Stintzing and Zwerger (1997) verified that model by calculating the coherence length ξ which can be regarded as the spatial extent of an electron pair, and the London penetration depth λ which is the thickness of the layer in which a magnetic field can enter the superconductor in its superconducting phase (Lifshitz and Pitaevskii 1980). For two frequently used ceramic superconductors (YBCO: Yttrium-Barium-Copper-Oxide, and BSCCO: Bismut-Strontium-Calcium-Copper-Oxide) the experimentally measured ratios of λ and ξ are 100 and 86, respectively (Stintzing and Zwerger 1997). Parameterising the superconductor model with values from experiments on YBCO and BSCCO led to ratios of 99 and 87, respectively, which is an extraordinary agreement between model prediction and observation – especially given that the model is extremely simple and ignores all of the atomic structure of these complex compounds.

As in the previous wildfire model, one can conclude that even crude models can lead to dynamics that almost perfectly agree with those of the real system, as long as they contain essential features – which in the case of the superconductor are basic quantum mechanical laws. Building models against the background of a hypothesis or a theory and testing the model's ability to reproduce observed characteristics of a real-world system allows us to assess the generality of the theory.

2.4 Understanding

While the previous three sections have dealt with the relationship between models and theory, the next four sections will be more applied –

with an increasing level of application. When exploring a system by way of modelling, the first step is usually the attempt to *understand* it (where theory can be helpful, as demonstrated in Sections 2.1 and 2.2). When a system is sufficiently understood it is possible to *explain* observed phenomena (which may be the motivation of the modelling exercise in the first place, but logically this comes after understanding). When the system has been understood and phenomena can be explained, it may be possible to use the model for making *predictions* of the system's future development. Since predictions may also include an assessment of the impacts of human interventions in the system, the model then may be used to *support decision-making* on how to use or manipulate the system to achieve particular objectives.

The delineation between understanding and explanation is rather fuzzy, and usually both activities are carried out simultaneously when analysing a model. The difference might be that the former activity is broader and includes an analysis of all the dynamics a model can produce while the latter is more focused on the analysis of observed or observable real-world phenomena.

An example that covers both activities but with a stronger focus on the general understanding of several systems is the recent analysis of global deforestation patterns by Taubert et al. (2018). Over the past few decades, rainforests have rapidly declined in many parts of the world. This decline seems to occur regardless of the political system of the respective country or the country's socioeconomic and natural conditions. The question addressed by Taubert and colleagues is whether there are differences in the spatial patterns of deforestation across the world. To answer this question the authors determined the current frequency distribution of the sizes of the forest fragments separately on three continents, the Americas, Africa and Asia-Australia, using satellite images with a resolution of 30 m by 30 m. As a first result it turned out that the sizes of the forest fragments on all continents range over several orders of magnitude. An entirely unexpected observation, however, was that the size distributions of the forest fragments are identical on all continents and obey a power law (cf. the distribution of wildfire sizes in Section 2.2), so that the frequency $f(A)$ of observing forest fragments of size A is given by

$$f \sim A^{-\beta} \tag{2.8}$$

with $\beta = 1.90$, 1.98 and 1.92 for the Americas, Africa and Asia-Australia, respectively. As demonstrated in the previous sections, such a similarity indicates a common underlying general principle that governs the dynamics.

A good candidate for such a general principle, as hypothesised by the authors, is percolation theory (Stauffer and Aharony 1994). Basically, in a percolation process (forested) pixels out of a fully covered region ('continent' with continuous forest) are removed at random. In the beginning of this process, the forest will still be largely continuous, only 'disturbed' by some empty (deforested) pixels amongst them. As the percolation process carries on and more and more pixels are removed, the previously continuous forest declines in extent and at some stage breaks up into two (mostly still very large) fragments. Carrying on further, these fragments break up, little by little, into smaller fragments, and so on. At some stage only a single large fragment is left, with a spatial extent similar to that of the entire region, plus some fragments of other sizes. One can show that, at this point, termed the 'percolation threshold', fragments of many different sizes coexist. As the percolation process carries on further, an increasingly smaller part of the region is covered by forest and the fragment sizes decline rapidly until only very small fragments are left.

The link between percolation theory and the fragmentation pattern observed by Taubert et al. (2018) is that at the percolation threshold the fragment size distribution obeys the power law of Eq. (2.8) with $\beta = 2.05$ (Taubert et al. 2018). The similarity between the simulated value of β and the observed values strongly suggests that – despite the local particularities of the three different continents – the rainforest loss all over the world can be described by a simple percolation process.

This insight even allows for some prediction and decision support. Next to the fragment size distribution, percolation theory predicts that, after the percolation threshold has been crossed, the larger fragments rapidly (i.e., in the form of a phase transition as described in the previous sections) disappear and only small fragments remain – a situation that has to be avoided by all means given its adverse consequences for the world's climate and biodiversity. To investigate how close the world's rainforests currently are to the percolation threshold, Taubert et al. (2018) compared observed forest fragment size distributions with those from the simulated percolation process and found that rainforests on all continents are already very close to the percolation threshold. Although this conclusion may not yet be regarded as concrete decision support (which will be covered in Section 2.7), it provides strong policy advice that deforestation in the tropics must slow down and cease as soon as possible, in order to avoid tipping over into a 'phase' with unacceptably high levels of fragmentation.

One reason why the world's rainforests keep declining despite wide-spread knowledge of their critical state and their global relevance is that they represent, in economic terms, a public good. Public goods are non-rivalrous and non-excludable. The former means that consumption of the good does not reduce its availability to others (in the case of a rainforest one should clarify that this holds only as long as the forest is present), while the latter means that individuals cannot be excluded from consuming the good (in rainforests such an exclusion is possible in principle but difficult in practice).

The main problem with public goods – including many environmental goods such as biodiversity or clean air – is that everybody can consume them without paying anything (or only a minimal price) to produce or preserve them (for more details, see Section 9.1). Why does this mismatch between benefits and costs more or less inevitably lead to the overuse of the public good? An explanation and deeper understanding of the public good problem is provided by game theory (for more detailed coverage, see Chapter 10). Game theory deals with the actions of (human and non-human) individuals and their mutual interactions. It often studies rather simple situations with only two interacting individuals ('players'), but for this very reason it allows us to derive very intuitive and instructive insights.

To provide an example, and taking up the problem of deforestation, consider two adjacent countries sharing a continuous cross-border rainforest. We assume that the use of (conserved) rainforest is non-rivalrous in that both countries can draw revenues from both portions of the forest, for example through tourism. We further assume that no country can exclude the other country from drawing tourist revenues from either portion of the forest. These two assumptions qualify the (conserved) rainforest as a public good. The dilemma with the conservation of this public good may be demonstrated by the following numerical example. Developing the rainforest to establish, for example, an industrial site earns for the developing country a revenue of 100 million dollars. If, in contrast, one country conserves its forest, *each* of the two countries (due to the non-rivalry and non-excludability) earns 80 million dollars. If both countries conserve their forests, each country earns $2 \cdot 80 = 160$ million dollars. Four different land-use patterns are possible: both countries develop, country A develops and B conserves, country A conserves and B develops, and both countries conserve. For each case the revenue (or 'pay-off' in the language of game theory) of country A (the result is the same for country B, since the game is symmetric) is given in Table 2.1.

Table 2.1 *Payoff for country A (in million dollars) as a function of each country's choice to conserve or to develop its forest (after lecture notes of S. Baumgärtner). The first and second summands in each entry are the contribution of country A and country B, respectively, to country A's pay-off.*

		Country B	
		Conserve	Develop
Country A	Conserve	80 + 80 = 160	80 + 0 = 80
	Develop	100 + 80 = 180	100 + 0 = 100

The public good problem is evident from the numbers in the pay-off matrix. Whatever country B does, country A always receives a higher pay-off by developing its rainforest. Or in other words, neither of the two countries has an incentive to conserve its forest. In the language of game theory, the 'stable state' in which both countries develop their forest is termed a Nash equilibrium in which any country that deviates from the equilibrium (in the example: by conserving rather than developing its forest) is worse off, and so the equilibrium of both countries developing is stable. The tragedy with this actuality is that if both countries decided to conserve their forests each of them would earn a revenue of 160 million dollars. But instead, they are stuck in the Nash equilibrium that is associated with a revenue of only 100 million dollars each.

To summarise the present section, two models have been presented that help understand causes of environmental degradation, the first dealing with the process and impacts of deforestation in tropical countries, the second one with reasons why it is difficult to stop this process. The latter raises the question if and how it is possible to leave the unfavourable Nash equilibrium and move into the favourable state of conservation, which will be addressed in Part III of this book.

2.5 Explanation

When the behaviour of a model has been sufficiently understood, the model can be used to explain real-world phenomena. The present section will present two models, an ecological and an economic one, designed to explain phenomena observed in the real world.

The first model addresses the ongoing debate about the cause of the so-called fairy circles occurring in arid grasslands, especially in Namibia.

The fairy circles denominate circular areas of bare soil with a diameter of a few meters, arranged as a regular grid within a matrix of vegetation. Two hypotheses for the explanation of fairy circles compete with each other with some intensity: that the fairy circles are caused by termite activity, or that they represent an emergent phenomenon from the interaction of rainfall, water dynamics and vegetation growth. The paper by Getzin et al. (2016) analyses the spatial pattern of fairy circles recently observed in Western Australia to support the second hypothesis with two arguments. First, the authors use spatial statistics to show the high regularity of the pattern: with little error all fairy circles in the study region are arranged on a regular hexagonal grid with a distance between neighbouring circles of about 10 m. No correlation could be observed in the study region between the presence of fairy circles and observed termite activity.

For their second argument, the authors adapt and parameterise a vegetation model from the literature to the conditions present in the study region, considering rainfall, water infiltration and evaporation rates, shapes of the plants' roots, and more. The model describes, through three coupled partial integro-differential equations, the spatio-temporal dynamics of three interacting quantities: vegetation biomass, surface water and soil water. It turns out that the parameterised model can reproduce the general features (grid structure, direct-neighbour distances, etc.) of the observed pattern with high precision. The authors conclude that pattern formation, caused by self-organisation that can be observed in many physical, chemical and biological systems, is the main cause behind the fairy circles. The authors' conclusion is further supported by the arguments already mentioned in Section 2.2: that a model or a theory gains credibility if it is applicable to a wide range of systems. While pattern formation had already been discussed as a plausible cause for the fairy circles in Namibia, the authors show that pattern formation is also able to explain fairy circles in a very different system with different environmental conditions.

The second example of a model being used for explanation is the so-called cobweb model of the famous pork cycle which was originally observed in European and US pig markets early in the twentieth century, but has since been used as a metaphor for cyclic dynamics in economic markets altogether. In those years the prices for pigs had been observed to oscillate significantly with an amplitude of about ±20 per cent and a period length of about five years (Hanau 1928). Authors in the 1920s and 1930s speculated that these oscillations could stem from a time delay between the observed pig price and the produced number of pigs. In an ideal economy, supply and demand for a good are in equilibrium and if, for whatever reason, supply falls below that equilibrium the market price for the good instantaneously increases, triggering an instantaneous response in the supply and/or the demand so that the market instantaneously moves back to its equilibrium.

Such an instantaneous response is often unrealistic. In the case of piggery, for example, a producer may observe a high pig price and decide to expand its production, but it takes time for stables to be built and pigs to be grown. By the time the pigs can be brought to the market, the pig price might have dropped again, causing an oversupply of pigs.

Following the notation of Leydold (1997), these dynamics can be cast into a simple model:

$$
\begin{aligned}
s_t &= -\alpha + \beta p_{t-1} \\
d_t &= \gamma - \delta p_t \\
d_t &= s_t
\end{aligned}
\qquad (2.9)
$$

The first equation of Eq. (2.9) demonstrates that the supply s_t of a good at some time step t is positively related to the price p_{t-1} of the good in the previous time step, which captures the mentioned time delay between the decision to produce a good (based on the current market price p_{t-1}) and the time when the good can be supplied. The second equation models that the demand d_t for the good is negatively related to the current market price p_t. The supply and demand functions are depicted for three combinations of the model parameters β and δ in Fig. 2.2a,c,e. The third equation of Eq. (2.9) shows that at any time t supply must equal demand (more details about demand and supply functions and about the concept of market equilibrium can be found in Section 9.1).

Equation (2.9) can be solved for the market price as a function of time:

$$
p_t = \bar{p} + (p_0 - \bar{p}) - \left(\frac{\beta}{\delta}\right)^t
\qquad (2.10)
$$

with

$$
\bar{p} = \frac{\alpha + \gamma}{\beta + \delta}.
\qquad (2.11)
$$

(Leydold 1997). For three combinations of the model parameters β and δ, Fig. 2.2b,d,f shows the temporal development of price p_t and quantity q_t. If $\beta < \delta$ (Fig. 2.2a,b), the oscillations diminish and the price converges to $\bar{p}$ of Eq. (2.11). For $\beta = \delta$ (Fig. 2.2c,d) and $\beta > \delta$ (Fig. 2.2e,f) the oscillations have a constant and an increasing amplitude, respectively.

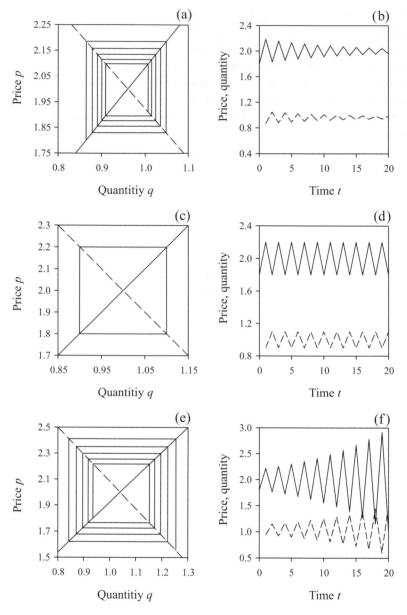

Figure 2.2 Dynamics of the cobweb model. Panels a,c,e: phase diagram with price p plotted as a function of quantity q for several time periods; the solid and dashed straight lines represent the supply and demand functions, respectively (Eq. 2.9). Panels b,d,f: price (solid line) and quantity (dashed line) plotted as functions of time t. Model parameters: $\beta = 0.48$, $\delta = 0.52$ (panels a,b), $\beta = 0.50$, $\delta = 0.50$ (panels c,d), $\beta = 0.52$, $\delta = 0.48$ (panels e,f); $\alpha = 0$, $\gamma = 2$.

The cause of the oscillations can be understood from the phase diagrams in Fig. 2.2a,c,e, in which the quantity of the good is plotted versus the good's market price. Considering Fig. 2.2a and starting with a price of $p = 1.8$, the supply in the next time period is given by $s(p = 1.8) = 0.86$ (lower left point in the 'cobweb'). The demand in that period must equal the supply (third equation of Eq. (2.9)), so the current demand is given by $d = s = 0.86$. This demand is associated with a market price of $p = 2.18$ (upper left point in the 'cobweb'). This price triggers production and a supply in the *next* period of $s(p = 2.18) = 1.05$ (upper right point in the 'cobweb'). In that next period supply must again equal demand, implying a market price of $p = 1.83$ (lower right point of the 'cobweb'), which triggers a new supply in the following period and the cycle starts anew. Whether the amplitude of the oscillations increases or decreases depends on the slopes β and δ of the supply and demand functions, as told by Eq. (2.10).

The market dynamics plotted in the phase diagrams of Fig. 2.2 look like a cobweb, giving the presented model its name. The cobweb model has initiated an interesting debate about the role of rational expectations of actors in an economy. One could wonder why a pig producer, having observed several cycles in the past (probably associated with substantial financial losses), is not able to guess that a currently high pig price may drop in the next couple of years and adapt their decision on the production level accordingly. On the other hand, the ability of human actors to predict the future should not be overestimated, and so the seemingly pessimistic assumption about the cognitive abilities of market participants in the cobweb model may be realistic in some instances. Nevertheless, in Section 12.5 I will explore how the assumption of better foresight can change the dynamics of the cobweb model.

To conclude, if models are able to reproduce observed patterns from real-world systems and if the mechanisms included in the model can be shown to be responsible for the pattern generated by the model, these mechanisms are likely to be also responsible for the observed pattern in the modelled real-world system. Of course, one can never prove that there exists no alternative mechanism that can produce the observed pattern equally well, but if a sufficient number of alternative mechanisms or hypotheses have been tested and could be discarded as possible explanations, one can be quite confident that the identified mechanisms are indeed responsible for the observed real-world pattern.

2.6 Prediction

With regard to prediction one may distinguish two different levels. The first level is the trivial one: that any model produces some output as a function of some inputs, and by this any model is able to predict how its output depends on the inputs. In this way, prediction is an inherent purpose of any model. The second level of prediction as a model purpose is whether the model is able to make *correct* predictions for a *real-world system*.

The first level of prediction, though trivial, can be very useful in model development and parameterisation and for the use of the model for explanation and theory development. A model can be used for these purposes by comparing its predictions with observed behaviour of the focal system – a technique that is called 'pattern-oriented modelling' (Grimm et al. 2005). A pattern is an aggregated property of a system that abstracts from the details but includes information on a higher system level. For instance, in an age-structured population where individuals of different ages coexist, it is often not necessary to know the age of each individual but is sufficient to know the age structure of the whole population, that is, the proportions of individuals in different age classes (e.g. below 1 year old, 1–2 years old, etc.). Alternatively, patterns may be of a spatial nature, such as the frequency distributions of fire sizes and forest fragment sizes in the papers by Zinck and Grimm (2009) and Taubert et al. (2018), respectively; or the spatial arrangement of the fairy circles in the paper by Getzin et al. (2016). Lastly, patterns may also be dynamic, such as the cycles of the pig prices observed by Hanau (1928) discussed in the Section 2.5.

These patterns are significant fingerprints of the system, and comparison of modelled patterns with their observed real-world counterparts can reveal important information. For instance, different model variants of a system may represent different hypotheses regarding the processes governing the system. Obviously the hypothesis behind the model structure that leads to the strongest agreement between modelled and observed patterns is most likely to be the true one.

In addition to theory and hypothesis testing, pattern-oriented modelling can be used to parameterise a model (Grimm et al. 2005). Often, some information about parameter values (e.g. population growth rates or the marginal production cost of some good) exists already at the time when the model is constructed, but this information may be subject to high uncertainty. By running the model for a large number of model parameter combinations one can explore whether the model output matches, according to some quantitative criteria, a set of observed patterns. Each pattern is used as a filter; parameter combinations that are not able to match this pattern are rejected. Only the reduced set of parameter combinations that allows the model to match all observed real-world patterns is then used for further model analyses.

Experience shows that the significance of the conclusions drawn from pattern-oriented modelling increases with the number and independence (e.g. spatial *and* temporal) of the patterns employed. Different techniques exist for pattern-oriented modelling, ranging from simple filtering (where maximum deviations between modelled and observed patterns are defined and all model parameter combinations that lead to smaller deviations are accepted) to Bayesian Monte Carlo Markov Chains where the model is simulated for a large number of repetitions with randomly but efficiently sampled parameter values until a distribution of likely parameter values emerges (Hartig et al. 2011).

The tuning of a model through pattern-oriented modelling is likely to improve the model's ability to generate precise predictions of the future development of the system. Still, the question remains how good the predictions of mechanistic models that consider the system processes through mathematical rules or equations are compared with those of statistical or so-called model-free forecasting approaches (such as regression analysis: e.g. Dobson (1983)) that make no assumption about the processes behind the observed spatio-temporal dynamics but 'simply' correlate data. That the answer to this question is not trivial has been shown by Peretti et al. (2013). Considering four different mechanistic ecological models, the authors chose certain values a_i for the model parameters, $i = 1, \ldots, N$, and simulated the dynamics of the model systems to produce time series of population sizes. The four models considered cover a wide range of model structures, such as spatial structure and age structure, and a wide range of possible system dynamics, such as chaos (cf. Section 4.3).

The time series then were used to make predictions of the future development of the four model systems in two manners. In the first approach, the models were fit, through Monte Carlo Markov Chain, to the time series to estimate the parameters of the mechanistic models. Due to the uncertainties in the model dynamics, the estimated parameter values a_i' differed from the true values a_i. The tuned models were then run with the estimated parameter values a_i' to simulate the future dynamics of the model systems. The outcome was compared with that of the true model with parameter values a_i (in fact, the continuation of the time series produced in the first place) and the deviation between predicted population sizes and 'true' population sizes was recorded.

In the second approach, several 'model-free' time series models that calculate current population size as a simple function (e.g. an average) of previous populations sizes were fitted to the produced time series to estimate the parameters of that function. Again, the time series models were used to predict the development of the model systems and the predicted dynamics were compared with the true ones.

The surprising result is that the time series models, though not containing any qualitative a priori information about the processes governing the four model systems, produced better predictions than the mechanistic models. And even more surprising, the error of the mechanistic model was largest when the model dynamics were chaotic – just where one could have expected that a complex mechanistic model that by design is able to produce chaotic dynamics would outperform at least the simple linear time series models considered. As the authors explain, the reason for this unexpected result is that the estimated parameter values a_i' in the mechanistic model reflected only the average behaviour of the model systems and in the simulations led to stable rather than (the true) chaotic dynamics.

Recalling the outline of pattern-oriented modelling, the only possibility to reverse the unexpected conclusion of Peretti et al. (2013) seems to be that the authors may not have used the patterns (i.e. the observed time series) as efficiently as possible. Perhaps the patterns represented by the time series were not independent enough, so the fitting of the mechanistic models was not able to capture the richness of the dynamics the models can produce. Nevertheless, there may be a trade-off between the two purposes of modelling – understanding and explanation on the one hand (for which mechanistic models are clearly more suitable than statistical or model-free forecasting approaches) and prediction on the other hand – and one should be clear about whether the main purpose of a modelling exercise is explanation or prediction (cf. Levins 1966 and Shmueli 2010).

2.7 Decision Support

If a model has an acceptable ability to predict the future development of a system it may also be used to model manipulations and influences by human actors on the system and assess the impacts of these influences with regard to certain management objectives. This ability of models to predict the impacts of human actions opens up the opportunity to incorporate models into decision support systems.

An example of such a model-based decision support system is the software DSS-Ecopay (Sturm et al. 2018). The motivation to develop this software was to assist policymakers in the development of agri-environmental schemes for the conservation of biodiversity. Agri-environmental schemes, also known as conservation payments or 'payments for environmental services' (cf. Section 9.2), involve payments to farmers who manage their land in a biodiversity-friendly manner. The software DSS-Ecopay focuses on grasslands and associated bird and butterfly species. The viability of these species has declined in recent decades due to an intensification of agriculture, such as heavy fertilisation, too frequent mowing of meadows and overgrazing of pastures, as well as mowing or grazing at the wrong time when eggs or immobile offspring are situated on the grassland.

To stop the loss of grassland biodiversity, alternative extensive management regimes would have to be established. However, since the requirements of grassland species – for example with regard to mowing times – strongly differ, there is no single grassland management regime that supports all species. Instead, a landscape mosaic is required that consists of grassland patches with different management regimes. For each alternative management regime a conservation agency such as an environmental ministry may offer a payment to compensate the farmers for the income losses that arise if they do not manage their grassland in a profit-maximising manner.

The design of such a payment scheme is highly complex, since a multitude of several hundred management regimes must be considered (it is not known a priori which ones are best and how many different ones are required) whose ecological benefits and economic costs differ spatially. Computer software can be very helpful here, since it can integrate all the required information and the spatial and temporal dependencies, which it would not be possible for an individual to do without computational help. The software DSS-Ecopay allows for two types of analysis: simulating existing or planned agri-environmental schemes to assess their overall economic costs and ecological benefits, and identifying cost-effective schemes that maximise an overall ecological benefit (e.g. the total habitat area available for a selected number of target species) for a given conservation budget.

Among other purposes, the software has been used to evaluate an existing agri-environmental scheme in the federal state of Saxony in Germany and to identify a cost-effective scheme that maximises the areas of target species for a budget of 11.1 million euros (Wätzold et al. 2016). The authors showed that substantial efficiency gains could be realised by modifying the current set of supported grassland management regimes and the associated payments to the farmers. By pointing to those features of the current agricultural scheme in Saxony that should be modified to increase the scheme's cost-effectiveness, the software provides helpful policy advice and can support decisions in agricultural and environmental policy.

While the optimisation within the cost-effectiveness analysis in DSS-Ecopay relies on ecological and economic models incorporated in the software, the employed optimisation algorithm, simulated annealing, may be regarded as a model by itself. Simulated annealing (Kirkpatrick et al. 1983) mimics a cooling process in metallurgy in which liquid material is cooled gradually to produce large crystals, while minimising defects in the crystal's atomic structure. Such an ideal crystalline structure is associated with a minimum level of the crystal's free energy (cf. Section 2.1), so an annealing process basically searches for the crystal structure that minimises the crystal's free energy.

In the application of simulated annealing for solving an optimisation problem, the structure of the crystal, that is, the location of the atoms, is identified as the space of decision alternatives (e.g. on which grassland patch to apply which management regime), and the free energy F stands for the variable to be minimised (or maximised if multiplied by -1). The optimisation process starts from a randomly chosen initial management alternative ('crystal structure') with some performance with regard to the conservation objective ('energy F_0'). A second management alternative that is similar to but slightly different from the initial one is generated randomly and its performance (F_1) is compared with that of the initial alternative. If it outperforms the initial one (leading to a lower energy $F_1 < F_0$) it is preferred; if in contrast it performs worse (has higher energy) it is not automatically rejected but still accepted with some probability. In the next step of the optimisation process a third management alternative is generated randomly, compared with the current one and again accepted or rejected. The process of generating new management alternatives and comparing them with the previous ones is carried out for a large number of iterations until the optimal alternative (with minimum possible energy) has been found.

With regard to the optimisation, accepting even worse alternatives with non-zero probability sounds odd at first but has the advantage of avoiding getting stuck in a local optimum (i.e. an alternative that locally outperforms all *similar* alternatives but differs from the global optimum, which outperforms *all possible* alternatives). *Always* rejecting the worse alternative would be like always walking down a mountain at the steepest slope in order to get to the deepest valley. However, while this path maximises the chance of finding *some* valley, it will obviously not necessarily lead into the *deepest* valley. Since in the beginning of the optimisation process the risk of walking towards a local but not global optimum is highest, the probability of accepting a worse alternative (the temperature in the crystal) is chosen rather high in the beginning.

An example is depicted in Fig. 2.3a where x values in the entire range between −1 and +2 are possible, although values on the right-hand side, especially around the minimum of $f(x)$ at $x = 1.4$, are most likely. In the course of the optimisation process the temperature is gradually reduced (according to some 'cooling schedule') and suboptimal x values (associated with large $f(x)$ as in Fig. 2.3) become increasingly unlikely during the search process. If the cooling is slow enough, x will be 'trapped' in the 'domain' of the global optimum (right-hand side of Fig. 2.3b). After this stage the risk of approaching only a local optimum (on the left-hand side of the figure) is negligible and the temperature can be reduced further, so larger values are not accepted at all anymore, and x is forced closer to the global optimum (Fig. 2.3c). When the temperature reaches zero value (not shown in Fig. 2.3), x finally converges to the global optimum.

While simulated annealing is applicable even to complex optimisation problems, its disadvantage is that there is no guarantee that it really converges to the global optimum and does not get stuck in a local optimum. In Fig. 2.3b, for example, there is no guarantee that x is really trapped on the right-hand side; one can only say that the probability of being trapped on the left-hand side is very small. Despite the risk of failing to find the global optimum, simulated annealing is used in many optimisation tools. In the field of conservation biology its most popular application is in the decision support software Marxan (Ball et al. 2009) that identifies the optimal selection of conservation sites with respect to a given conservation objective. The Marxan software is very flexible and allows for considering many different types of management alternatives, objective functions and optimisation constraints.

To summarise, if a model contains the relevant features of a real system so that it can predict the future of the system dynamics with acceptable

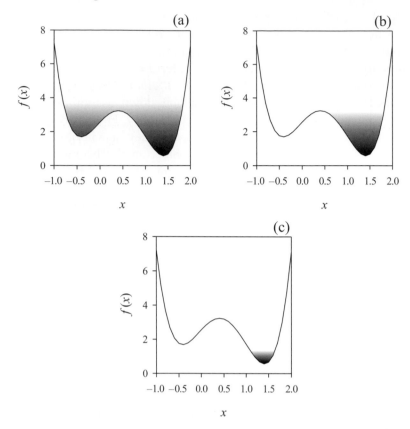

Figure 2.3 Example of a function $f(x)$ with a local and a global minimum. The colour represents the likelihood of an x with associated $f(x)$ being accepted in the course of the optimisation process, so that an x associated with a large $f(x)$ is less likely to be accepted (light colour) than an x associated with a small $f(x)$ (dark colour).

accuracy, then the model can be used to test management actions and assess their impacts on the system. In this manner the model supports decision-making, and many decision support systems consist of a model equipped with a user-friendly interface for convenient input and output of information.

2.8 Communication

While in the previous sections models were used to analyse systems (with the purpose of understanding them, predicting their behaviour, etc.), the following two sections present examples in which models are used to

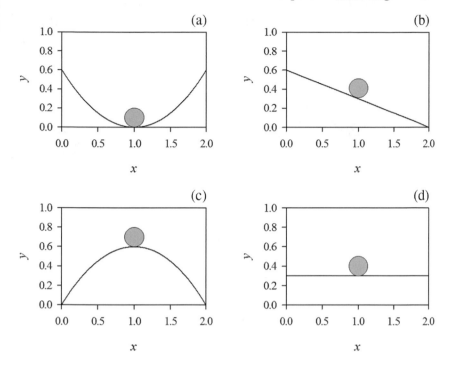

Figure 2.4 Depiction of different types of static (in-) equilibria.

communicate existing knowledge to other people. In the present section communication takes place in a classical manner from an active sender to a comparatively passive recipient, while in Section 2.9 the recipient is an active learner.

Sometimes a system or a process is well understood and predictions can be made even without a model. In such a case a model can still be useful to communicate the available knowledge. The map in Fig. 1.1 is an example. It does not generate any new knowledge but simply transmits existing knowledge from the producer of the map to the user.

A slightly more sophisticated communication model was used to explain the process of simulated annealing. The x-axis in Fig. 2.4 could represent a simplified one-dimensional representation of a possibly high-dimensional space of possible decisions and $f(x)$ could represent a complex objective function. Referring to the annealing process in metallurgy which gave the optimisation method its name, the x-axis represents all possible structures of the crystal and $f(x)$ the crystal's free energy. The shaded areas represent the thermal movements of the atoms,

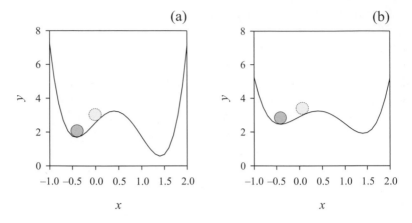

Figure 2.5 Depiction of multiple equilibria.

and the different intensities of the colour represent the likelihood of observing a particular structure in the crystal with the associated free energy.

At the minimum of $f(x)$ the crystal has minimum energy and assumes some equilibrium state, which in Section 2.7 was also compared to residing in a valley within a mountain area. This metaphor is often used to depict the properties of static equilibria (Fig. 2.4). Its position in the depression (panel a) minimises the ball's energy, and since physical bodies attempt to assume a state of lowest energy, this position is stable. In contrast, the position of the ball in panel Fig. 2.4b is unstable and the ball will roll down to the right. The position in Fig. 2.4c is an unstable equilibrium and the one in Fig. 2.4d is indifferent.

These metaphors can also be used to depict more complex forms of stability. Figure 2.5a shows a situation where the dark grey ball is in a stable position. However, if due to some disturbance it is moved away from its position – for example, to the light grey position – there is a risk that the ball will cross the ridge towards the right and roll down into the deeper depression to the right. The ridge thus depicts a tipping point, and a system may move to a very different state when a tipping point is crossed. If that alternative state is undesirable it is important to maintain the system in the original equilibrium (on the left in Fig. 2.5a). Obviously, the deeper the depression on the left-hand side, the more resilient the system will be to disturbances; and the stronger the disturbance, the higher the risk will be that the system will cross the tipping point towards the right-hand side.

A change in the system may be represented not only by an elongation of the ball from the dark grey to the light grey position but also by a distortion of the stability surface (Fig. 2.5b). If the ridge between the two depressions is flattened, the probability of crossing over from the left to the right equilibrium is increased. These (briefly touched on) issues of local stability, disturbance and tipping points are core elements of the concept of resilience (Walker et al. 2004), which is highly non-trivial but can be communicated very intuitively by graphs such as the one in Fig. 2.5.

Graphical metaphors may also be used to depict dynamics, such as the change of a stock due to flows. An example is the carbon stock in the atmosphere. An argument sometimes heard from 'climate sceptics' is that the increase in CO_2 emissions by humans is small compared with natural emissions. And in fact, according to Denman et al. (2007, Fig. 7.3) the annual emissions from the use of fossil fuels in the 1990s amounted to 6.4 gigatonnes of carbon (GtC) (5.4 GtC/yr if human land use is considered), which is small compared with the natural emissions of 190.2 GtC/yr. What is missing in this equation is, of course, that the earth's land and water surface at the same time absorbs 190 GtC/yr, so that natural emissions and absorption are almost exactly balanced, with a small net increase in the atmospheric carbon stock of 0.2GtC/yr.

To scientists, especially those with some mathematical background, it is obvious that this net increase in carbon flow, which due to human activities multiplies by a factor of 28–5.6 GtC/yr, is what really counts. To a lay person, however, it may be more instructive to imagine a water tank collecting rainwater at a given rate, say 190 litres per month. The same amount might be taken by the gardener every month to water the plants. Inflow and outflow of water are balanced and there is no risk the tank will ever become empty (an overflow would more closely relate to the carbon issue but would probably be less of a problem to the gardener). However, a comparatively small decline in the monthly rainfall by 5.4 litres (about 3 per cent) will tip the balance, so the tank will lose those 5.4 litres every month and will soon be empty.

The examples in this section demonstrate that a model does not necessarily need to be a highly complex mathematical construct, but can also be an image or a metaphor. Often it is useful to cast the insights drawn from a more complex model in a simple verbal or graphical model that is easier to understand than the original model but still captures its essence.

2.9 Education

Practically any model that produces some output from certain inputs can be used for education if it is equipped with a user-friendly interface through which a user can input values of model parameters and observe a resulting output, such as the dynamics of a species population or the market price of a good. As an example, Ulbrich et al. (2008) developed a software tool for investigating the cost-effectiveness of grassland mowing regimes (determined by frequency and timing of mowing events), so that the survival of an endangered butterfly species is maximised for a given conservation budget. The user can select the conservation budget and one of 112 different mowing regimes and simulate and evaluate the dynamics of the species.

Psychological research emphasises the role of active learning in which the learner does not simply 'consume' the teaching material but takes an active role in the learning process. The above-mentioned software already contains some active elements by assigning the user the role of an experimenter who can manipulate the parameters of the model system and observe the system's response. However, the user can be even more actively involved if they become an integral part of the model system. This is possible if the model includes human actors like in an agent-based model (cf. Chapter 12). Such agent-based models can be rather easily expanded into computer games by replacing the decision module that models the decisions of the agents with a user interface through which players can select the agent's decisions – effectively becoming agents themselves.

Learning about the system is achieved here by playing the system. This approach was taken, for example, by Hartig et al. (2010), who developed a multi-player computer game to demonstrate the functioning of conservation offsets (cf. Sections 9.4 and 13.2) to students during a science fair. Each student took the role of a landowner who could decide between conserving land and generating credits for sale on the market, or buying credits on the market and carrying out agriculture.

A more comprehensive view on biodiversity conservation and the sustainable use of land is conveyed by the online game LandYOUs by Schulze et al. (2015). In LandYOUs the player takes the role of a governor or land manager who manages a model region over ten time periods. In each period the player can invest in several land-use measures such as agriculture, afforestation or settlement. These investment decisions have an impact on the ecological and socioeconomic state of the model region in the next period of the game dynamics, which is communicated to the player by a set of indicators. In that next period the player can respond to the indicator values by appropriately adapting the investments into the above-mentioned land-use measures. During and at the end of the game the player receives a score, calculated from the ecological and socioeconomic indicators, that measures the level of sustainability achieved in the model region. Through this response the player can learn about sustainability and how it can be achieved through appropriate land use in a complex ecological-socioeconomic system.

2.10 Integration of Knowledge

The following two sections outline two technical roles of models in scientific research. The first is the integration of knowledge and the second is mediation between scales. A prerequisite of a model to fulfil one or more of the purposes outlined in the previous sections is that the model integrates qualitative and quantitative information from different sources in a structured and logical manner. Thus, a model represents a compilation of data, like an encyclopaedia, but in addition contains all the necessary logical links between these data to draw conclusions that are 'more than the sum' of the individual 'data bits'. This integration of data is usually carried out through mathematics, be it mathematical equations (e.g. to describe how the supply of a good depends on its price: Eq. (2.9)) or rules (e.g., 'a forest stand burns with a certain probability if at least one of its neighbouring stands burns': Zinck and Grimm 2009 in Section 2.2).

In the construction of an applied model for solving a specific biodiversity-conservation problem the first step may be to compile all the available (and relevant) knowledge. The software DSS-Ecopay (Section 2.7), for example, contains more or less everything that is known about the reproductive cycle (egg deposition times, breeding times, environmental breeding requirements, etc.) of the considered species. This data is then combined with the dynamics of land-use measures to assess how strongly the land-use measures interfere with the species' reproductive cycles. Since this may depend on environmental conditions such as altitude or soil quality, geographical data on these quantities had to be gathered and integrated as well.

Another example worth mentioning here is the BEEHAVE model by Becher et al. (2014), the first model of honeybee colonies that combines knowledge about processes within a hive with foraging and availability of flowering plants in the landscape. Earlier models focused either on within-hive dynamics or on foraging. The model includes the bees' life cycle in the colony (through a stage-based submodel: cf. Section 7.1), the dynamics of parasites and pathogens, the foraging behaviour of the bees (through an individual-based submodel: cf. Section 7.2) and the interactions between these processes. Thus the model not only integrates data scattered across various journal papers, scientific reports and other sources, but integrates it to generate added value in terms of understanding and prediction, because it takes into account the interaction between forage availability in the landscape (Horn et al. 2016) and colony dynamics.

2.11 Mediation between Scales

A valuable feature of models, along with their ability to integrate information, is that they can mediate between scales. By 'mediate' I mean that models allow for considering processes across spatial, temporal and organisational scales, and transferring information between these different scales. When investigating an ecological or an economic system empirically, a typical problem is that, due to resource constraints, data can be gathered only for a short time period and for small spatial areas. For understanding a system or supporting decision-making, however, often the long-term dynamics of the system need to be known, and usually not only on small but on several spatial scales. Models can be enormously useful in the extrapolation from small to larger scales. The model of Zinck and Grimm (2009), for example, uses only local information about the vegetation dynamics within individual forest stands and the interaction between neighbouring forest stands to model the fire and vegetation dynamics of the whole forest over long timescales – and to determine properties such as the fire size distribution on a large spatial scale.

Similar extrapolations can be carried out on the temporal dimension. O'Hara et al. (2002), for example, fitted a metapopulation model that describes the dynamics of multiple interacting local populations in a landscape (see Section 6.1) to a time series of ten years. For each of the ten years, information exists about which of the habitat patches in the study area are occupied by the focal species and which are empty. Through Bayesian Monte Carlo Markov Chain (cf. Section 2.6), the authors fitted the five parameters of the metapopulation model. After the model had been fitted to the ten-year time series, the authors used it to simulate the metapopulation dynamics for a much longer time span of

150 years. This allowed for predictions about the long-term viability of the species, which could not be generated initially from the ten years of data.

Models can mediate not only between different spatial and temporal scales but also between organisational levels. Fahse et al. (1998) presented a model study on the dynamics of nomadic larks in the Karoo in South Africa. The model considers quite complicated behaviour by the birds, which organise themselves into flocks searching for intermittent breeding areas. The behaviour of the birds was captured by an individual-based model (cf. Section 7.2) that describes the fate of each individual bird from birth to death, including movement in the landscape and reproduction. It turns out that the individual-based model is too complex to systematically run and analyse it over many years. To make the model more useable the authors employed the method of separation of timescales (Haken 1991), which simplifies the analysis of complex systems that exhibit dynamics operating on very different timescales. In the case of the nomadic larks, for example, the timescale of the individual birds' behaviours is much faster than that of the changes in the total population size. Therefore, for modelling the change in the total population size it is not necessary to know all the details of the individuals and their behavioural dynamics, but only the parameters of classical population ecology: birth and death rates (cf. Section 4.1), which can be calculated from the birds' individual behaviour.

These calculations extrapolate not only from short to long timescales, as in the previous example, but also from the low organisational level of the individual to the higher organisational level of the population. The parameters (birth and death rates) describing the dynamics on the population level emerge from the dynamics on the level of the individuals. The approach presented by Fahse et al. (1998) also opens up opportunities for generalisation, since it shows that it is possible to abstract from the details on the individual level and consider only the dynamics on the population level.

2.12 The Trade-Off between Generality and Specificity

The two final sections address two broad dichotomies between model purposes: first, whether a model aims at generality or at specificity, and second, whether it is designed to derive positive or normative conclusions. Models can generally be classified into having two contrasting purposes: to provide general insights on an abstract level or specific insights into a specific system (Baumgärtner et al. 2008). This purpose is usually reflected in the complexity and level of detail considered in the model. Simple models are often more suitable than complex models for generating general insights but are hardly applicable to specific systems without large error. Complex models, in contrast, can take all the details of the focal system into account but are often of limited use in the generation of general knowledge and understanding.

Given this trade-off between generality and specificity and between simplicity and complexity, a promising approach for the analysis of complex systems is to build a suite of models of varying complexity. Two approaches are possible. The first option is to start from a complex model that agrees very well with empirical observations in the focal system and simplify it step by step, and in each step check whether and how much of the agreement is lost. In this approach, which has been dubbed 'robustness analysis' by Grimm and Berger (2016), the task is to 'break' the model, that is, to explore which processes and parameter settings are essential for the model to be able to reproduce observed patterns. The second option is to start from a simple model and gradually add more detail; this approach was adopted by Thulke (1999), who developed a very simple epidemiological model for the spread of rabies. While that model was able to capture the general spatio-temporal pattern of the spread of the disease, it disagreed with more detailed observations in a particular real landscape. Therefore the authors replaced, step by step, some functional relationships of the simple model with more detailed submodels, and in so doing succeeded in gradually improving the precision of their model. These approaches of increasing or decreasing the level of complexity of a model are very effective for detecting those model features that are necessary to produce the patterns observed in real systems, and to develop an understanding of the similarities and differences between systems.

2.13 Positive versus Normative Analysis

A final distinction between the purposes of modelling that should be mentioned is whether a model is used to derive positive or normative statements. The distinction between positive and normative statements is primarily found in philosophy and the social sciences. A positive statement is descriptive and concerns what was, is or will be, while a normative statement concerns what should be. By this, a normative statement depends on assumed ethical values. Whether a certain policy or land-use

strategy is judged superior to another depends not only on its performance with regard to its ecological, economic and social impacts but also
on the importance assigned to these impacts, which reflect the ethical
values of the policymaker(s) or the people they represent.

Most of the previous modelling examples, especially the ecological
ones, are of a positive nature, since they investigate systems dynamics and
the relationships between causes and consequences – without judging
whether a certain consequence is desirable or not. Among the previous
examples, normative model studies are found mainly in Section 2.7 on
decision support. Decisions (at least if they are rational) always rely on
values. In order to decide whether a conservation policy is cost-effective,
that is, maximises biodiversity for a given budget, one has to specify
which species to target and how to weight their well-beings. Even the
choice of cost-effectiveness as a performance criterion involves the normative judgement that wasting resources is unethical, or at least
undesirable.

In the context of ecological-economic modelling, normative and
positive analysis is reflected in two tasks. For a positive analysis the
ecological-economic system is simulated to explore how it evolves and
how its dynamics are affected by model assumptions and constraints.
A normative analysis usually involves some sort of optimisation. An
analysis of cost-effectiveness, for example, requires that an objective
function (e.g. number of viable species) is maximised under given constraints (e.g. a limited budget). However, in the practice of ecological-
economic modelling there is often a mixture of positive and normative
analysis, because often an ecological-economic analysis does not end
with the identification of a cost-effective policy or strategy, but may be
followed by an exploration of the factors that determine the level of the
policy's cost-effectiveness. That again is a positive analysis.

To conclude this chapter on model purposes, a few references may be
recommended that deal with the role of modelling, what models are and
can do, and what they are not and cannot do: Starfield et al. (1990),
Starfield (1997), Getz (1998) and Nicholson et al. (2002).

3 · *Typical Model Features*

Models can be classified not only according to their purpose but also according to certain features and structural properties. These features should reflect the focal system and the model's purpose. For instance, to analyse the process of rainforest loss and fragmentation (Section 2.4) the model should take account of spatial structure; in order to model the pork cycle (Section 2.5) the model should be dynamic and contain a feedback loop, such that the price of the good affects its quantity at some later time, which then affects the price, and so on. In the present chapter, I will focus on five model features: spatial structure, dynamics, stochasticity, individual variability and feedback loops. The relevance of these features and the consequences of neglecting them will be demonstrated through a number of examples from the literature.

3.1 Spatial Structure

In many real-world ecological and economic systems, spatial structure plays an important role. In Drechsler and colleagues' consideration of space, four types of model may be distinguished (Drechsler et al. 2007): spatially implicit models, spatially differentiated models, spatially explicit models and spatially differentiated and explicit models. *Spatially implicit* models contain processes that reflect the spatial structure of the focal system but are non-spatial in the model structure. For instance, a landscape may be spatially structured into agricultural land parcels or habitat patches, but if all these have identical properties, and if the system dynamics do not depend on the distances between these spatial entities, the model dynamics will essentially be non-spatial. In *spatially differentiated* models the system dynamics are still independent of the distances between the spatial entities but the properties of spatial entities may differ, as does, for example, the agricultural productivity of a piece of land or the quality of habitat for a given species. In *spatially explicit* models, in contrast, the spatial entities have identical properties but the system dynamics depend on the distances between the spatial entities. And lastly, in *spatially differentiated and explicit* models the spatial entities differ in their properties and the system dynamics depend on the distances between the spatial entities.

An example of a spatially implicit model is the metapopulation model by Levins (1969). Originally developed to address questions in the field of pest control, it soon found its way into conservation biology and stimulated an entire research field: metapopulation ecology (Hanski 1999; see Section 6.1). A metapopulation is an ensemble of local populations, each of which inhabits a patch of suitable habitat. Local populations can occasionally become extinct due to demographic or environmental stochasticity (see Section 3.3 and Chapter 5), but empty habitat patches can be colonised by other local populations. In this way, the metapopulation can persist as a whole despite the occasional extinction of local populations.

Levins (1969) considered an ensemble of S habitat patches of identical size. Local populations become extinct at a rate (proportion per time unit) e, and each local population emits individuals at a rate c to colonise other habitat patches. Although a metapopulation consisting of a number of local populations is obviously spatially structured, its dynamics can be described by a non-spatial model if one assumes that an emigrant from a local population can reach any of the other habitat patches with equal probability, independent of the distances between the source patch and the other habitat patches. Under this assumption the dynamics of local extinction and colonisation, which are essentially represented by the dynamics of the number of local populations in the landscape, can be described by a simple (non-spatial) ordinary differential equation (Levins 1969; see Section 6.1). The information about the spatial structure is implicit, contained in the parameters e and c, such that larger habitat patches are likely to imply a lower local extinction rate e, and higher distances among the habitat patches imply a lower colonisation rate c.

In spatially differentiated models, spatial entities may have different properties. To make Levins' model spatially differentiated one could, for example, assume that the patches differ in size, which would imply that it is important to know which patch is currently occupied by the species and which is empty. However, if emigrants can reach any patch, the spatial locations of the patches do not play a role. Ando et al. (1998) offer an example of a spatially differentiated model analysis. They deal with the problem of selecting sites for conservation to protect the species at that location. The site selection should be cost-effective, so that the number of protected species is maximised in the focal region for a given budget (or alternatively, the budget required to protect a given number of species is minimised). In Ando et al. (1998) the sites considered cover all counties in the United States (about 3,000), and the focus is on the protection of about 900 species. While the properties of the counties differ, the model analysis does not explicitly consider the locations of the counties in the country.

In their 1998 paper, Ando et al. respond to an earlier paper that states that only a few key sites would suffice to protect endangered species 'with great efficacy'. What the authors of the earlier paper ignored, and what Ando and colleagues point out, is the spatial differentiation in the cost of conserving a site, because land prices strongly differ among the sites. To demonstrate the importance of their argument, Ando and colleagues compare two processes of site selection. In the first process they ignore the spatial cost differentiation among the sites and the budget is defined only by the number of conserved sites. The authors then calculate the financial costs of conserving the selected sites, taking the information on land prices into account.

In the second selection process, the authors consider the spatially differentiated land prices right away and again determine the cost-effective selection of sites. It turns out that the total cost associated with the sites selected in the second process is about 30 per cent of that associated with the sites selected in the first process, meaning that taking the spatial differentiation of the land prices into account triples the level of cost-effectiveness.

The third type of spatial model is spatially explicit models in which the interaction between entities depends on the distances between the entities, but the processes do not explicitly depend on the spatial location (no spatial differentiation). One of the first models of this kind is Conway's famous Game of Life (Gardner 1970). This considers a simple square grid of cells, each of which may contain an individual (or population) of a species or be empty. Like Levins' (1969) model, occupied cells can become empty while empty cells can become occupied. The transition rules are the same for all cells (so the model is not spatially differentiated), but the transition of a cell between its two possible states depends on the number of occupied cells in the cell's neighbourhood. In particular, an occupied cell remains occupied only if the number of occupied neighbours is within some interval, and an empty cell becomes occupied only if it is surrounded by a particular number of occupied neighbours (Section 6.2). Since the transition of a cell from occupied to empty or vice versa depends only on the number of occupied cells in the immediate neighbourhood, the interaction between different cells is distance-dependent, and in this respect Conway's model is spatially explicit. Consequently, and in contrast to the spatially differentiated models, it is important to know exactly where the occupied and the empty cells are located. To demonstrate the relevance of such spatial explicitness, in Section 6.2 I will modify Conway's Game of Life by assuming long-range interactions between the grid cells. The dynamics produced by that modified model turn out to be very different from those produced by Conway's original model.

Combining spatial differentiation with spatial explicitness leads to the fourth type of spatial model: spatially differentiated and explicit models. Conway's model, for example, could be made spatially explicit by introducing different habitat qualities in the cells. Some cells, for example, might require more occupied neighbours than other cells to change from empty to occupied. An example of a spatially differentiated and spatially explicit model is provided by Nalle et al. (2004). This model was used to investigate the cost-effectiveness of various spatio-temporal forest management strategies to conserve two forest species in a forest in Oregon, in the United States. The model consists of an economic and an ecological module. The economic module calculates the costs and revenues of different management strategies that can be applied to a forest stand. This module is spatially differentiated, since the revenues depend on the location of the stand in the study region. The ecological module is spatially differentiated because the quality of a forest stand for reproduction depends on the stand's age, and spatially explicit because the movement of species individuals between forest stands depends on the distances between the stands.

3.2 Dynamics

Most of the ecological and economic models found in the field of biodiversity conservation are dynamic (cf. Tables 15.1–15.3 in Section 15.2). Notable exceptions are, for example, the large number of reserve selection studies, such as the above-mentioned paper by Ando et al. (1998). Dynamics were considered implicitly by Ando and Mallory (2012), who applied modern portfolio theory to select reserve sites to maximise a joint biodiversity conservation objective in the face of uncertain climate change. The study used external model forecasts of the ecological values of potential conservation sites in a region in North America. These forecasts were derived from climate and ecological models and provided means, variances and co-variances of the sites' ecological values. Although those models are dynamic, the site selection problem itself was static.

Dynamic models have been used to analyse the effectiveness and cost-effectiveness of taxes and subsidies to penalise biodiversity-harming land use and reward biodiversity-friendly land use, respectively (Barraquand and Martinet (2011) for a stylised landscape, and Mouysset et al. (2011) for France). The cited models are dynamic in both their ecological and economic dimensions. Economic dynamics are induced, for example through imposed temporal changes in the levels of taxes and subsidies and the market prices for crops. Since the landowners are assumed to maximise their profit, the changes in economic drivers imply changes in land use. Land that is managed in a biodiversity-friendly manner provides habitat for species whose dynamics are described by classical population dynamic models. As a dynamic output, the models produce trajectories of land-use types, agricultural income and species abundance.

While in the models cited the dynamics are largely driven by exogenous economic drivers, the consideration of dynamics may be important even if these drivers are largely constant in time, as the study by Lapola et al. (2010) demonstrates. The authors analyse the impact of biofuel production targets on land use in Brazil. In a static setting one would expect that these targets lead to deforestation of the Brazilian rainforest in order to provide land for biofuel crops such as soybean and sugarcane. In the dynamic model analysis by Lapola et al. this direct land-use change is indeed observed, but it is smaller than the so-called indirect land-use change which encompasses a two-step displacement: rangeland for cattle production is displaced by biofuel cropland, and rainforest is logged in turn to provide new rangeland and compensate for the loss caused by the expansion of biofuel cropland. Naïvely, one might argue that on net the loss of rainforest would be rather similar – whether it occurs directly or indirectly – but politically and for the design of environmental policies it is important to understand whether land-use change is direct or indirect. Only a dynamic model of the kind used by Lapola et al. (2010) is able to detect this difference and assess how much of the observed land-use change is direct and how much is indirect.

3.3 Stochasticity

Whether the world is inherently deterministic or stochastic is a difficult question that is far beyond the scope of the present book. Nevertheless, even if it is deterministic many systems are so complex that their dynamics appear stochastic (Section 5.1). The extinction of species populations, for example, is affected by many unpredictable factors such as weather and the abundance of food or predators. Therefore, it is generally plausible to assume, for example, that the extinction rates of the local populations in Levins' (1969) metapopulation model (Section 3.1) are stochastic rates, that is, they represent the *probability* of a local population becoming extinct within some time unit. A similar point can be made about the colonisation rate in that model. As a consequence, the number of habitat patches occupied by local populations does not assume a fixed 'equilibrium' value (cf. Section 6.1) but fluctuates around it – and if the fluctuations are strong, the metapopulation can go extinct even if the equilibrium number of local populations is non-zero.

Thus, in a stochastic world concepts such as equilibrium lose their relevance, and statistical measures such as persistence time are more appropriate (Chapter 5). In models, stochasticity is represented via so-called pseudo random numbers which mimic real stochastic processes. Whenever, for a certain model purpose, variation or variability is believed to have important consequences, but we do not want, or are not able, to represent this variation mechanistically, we implement stochasticity. For example, whether a new canopy gap of a beech forest is taken over by a young tree within the gap or a neighbouring canopy tree depends on many factors, which we do not know in detail, and also do not need to know. Rather, it is sufficient to say that the chances are 50:50 (Rademacher et al. 2004).

The fact that the neglect of stochasticity can lead to wrong model predictions has been demonstrated by Frank (2005). The author compared two measures of metapopulation persistence. The first is the so-called metapopulation capacity, a deterministic measure that considers a metapopulation as persistent if in equilibrium all habitat patches have a positive probability of being occupied (Hanski and Ovaskainen 2000; Section 6.1). The metapopulation capacity can be interpreted as 'the average colonisation ability of the local populations in the case where only one of the patches is occupied' (Frank 2005, p. 376). The second measure of persistence is the probability of the metapopulation surviving over a given time horizon, which by definition is stochastic (Section 5.3). The survival probability is determined by the initial pattern of occupied patches and a quantity that is independent of the initial condition: the mean metapopulation life time (Frank 2005).

Comparing the two persistence measures on different networks with habitat patches of different sizes and pairwise distances, Frank (2005) showed that for weak stochasticity in the colonisation and extinction rates, both measures lead to very similar results. In contrast, in the case of moderate or strong stochasticity, the two measures lead to strongly diverging conclusions, for example with regard to the critical patch sizes required to deliver some target level of metapopulation persistence. In particular, the stochastic metapopulation survival probability was maximised by habitat networks with more evenly sized patches, compared with the metapopulation capacity which was maximised through more unevenly sized patches. The result that under strong stochasticity the total habitat area should be allocated rather evenly among all patches and not concentrated on a few patches relates to the well-known rule not to put all of one's eggs into one basket.

3.4 Individual Variability

Practically all ecological and economic systems exhibit variability or heterogeneity of some kind. An example is the above-mentioned spatial structure. The economic cost of conserving a land parcel for biodiversity often varies in space (Ando et al. 1998), and the suitability of a forest stand for a bird species may depend on the local management regime (Nalle et al. 2004). As a consequence of this space-dependent economic variability, some land parcels may be conserved under a particular policy and others may not, and the space-dependent ecological variability may imply that some habitat patches are occupied by a species and others are not.

In addition to this space-dependent variability, variability may be observed over time, or it may occur because the individuals (plants, animals or humans) differ, for example, with regard to their age, sex, health, nutritional status and other attributes. Further sources of variability are differences in the behaviour of individuals, such as memory, perception, learning ability, interaction with other individuals and so forth (cf. Section 7.2). Humans, in addition, may differ in their preferences for certain factors that determine human welfare, such as health, environmental quality and presence of iconic species (cf. Chapter 12).

Differences in these characteristics are likely to affect the dynamics of the ecological or economic system. If, for instance, age affects the reproduction and survival of individuals in a population, the age distribution in the population is likely to affect the population dynamics. This is important, for example, in the design of translocation strategies where individuals are taken from a source population to establish a new population elsewhere and expand the spatial range of the species. As Todd and Lintermans (2015) show, translocation as a conservation concept can involve trade-offs. Considering an endangered fish population, the authors argue that the removal of adults from the source population may negatively impact the source population, so it might be better to translocate subadults. Analysis of the authors' age-based model, however, reveals that a viable new population will be established from subadults only if these are introduced at rather high numbers – in particular at much higher numbers than would be required if the new population was established from adults.

As a second example, Skonhoft et al. (2012) analyse a model of a harvested age-structured fish population that consists of old and young adults. Two fisheries or fishing agents are considered, each of which targets one of the two age classes but also affects the respective other age class through by-catch. The authors show that differing properties of the two age classes, as well as assumptions about the level of information available to the fisheries, determine the sustainability of a harvesting policy.

3.5 Feedback Loops

Feedback loops arise if one component of a dynamic system affects another and vice versa, which usually leads to oscillations in the system dynamics. An example of a feedback loop has already been presented in Section 2.5 where the current price of a good affects the supplied quantity at the next time step, which affects the price in that time step, which affects the quantity in the following time step, and so on. That example already provides an important classification of feedback loops: that they can be positive or negative. In the case of a positive feedback, a small deviation from equilibrium in one system component induces a larger deviation in the other system component, which in turn induces an even larger deviation in the former component, and so on, so that the oscillations in the system dynamics grow (cf. Fig. 2.2e,f). In the case of a negative feedback, larger deviations in one system component lead to smaller deviations in the other, and so on, so that the oscillations decline (cf. Fig. 2.2a,b).

Perhaps the oldest model that involves feedback loops is the Lotka–Volterra model (named after the mathematicians Alfred J. Lotka

[1880–1949] and Vito Volterra [1860–1940]) which explains how the interaction between a predator and a prey species (e.g. fox and hare) leads to oscillations in the population dynamics of the two species. The model assumes that in the absence of the predator the growth of the prey population N (in particular, the time derivative dN/dt: see Section 4.1) is proportional to $N - \alpha N^2$, where the first term means that a larger population grows faster (in absolute terms) than a smaller population, while the second term considers that when the population becomes 'very' large, growth declines and can even become negative (Section 4.2). If, in addition, predators are present at abundance P, the growth of the prey population is further reduced, and that reduction is proportional to NP, which considers that the number of predation events increases with the number of prey and the number of predators. Altogether, the growth equation for the prey population reads

$$\frac{dN}{dt} = rN(1 - \alpha N) - \beta NP, \qquad (3.1)$$

where r is the per capita population growth rate at very small population sizes N and in the absence of predation ($\beta = 0$), and β measures the impact of a predator individual on the prey population. Predation naturally benefits the predator population, which grows proportionally to NP, while in the absence of prey the predator population declines from starvation:

$$\frac{dP}{dt} = \gamma NP - \delta P, \qquad (3.2)$$

where γ measures how strongly the predators can 'transform' predation into population growth and δ is the rate by which the predator population declines in the absence of prey.

Figure 3.1 shows examples of population trajectories obtained from Eqs (3.1) and (3.2). One can see that, depending on α, the feedbacks may be positive (with increasing amplitudes) or negative (with decreasing amplitudes and convergence to a stable point). A second, more subtle observation is that the prey population always precedes the predator population in time: if, for example, the prey population is currently large, the predator population has ample food and grows fast. This growth is associated with an increase in predation pressure, so the prey population declines. Once the prey population has declined substantially, the predators lack food and the predator population declines, too. This in turn reduces predation pressure and allows the prey population to

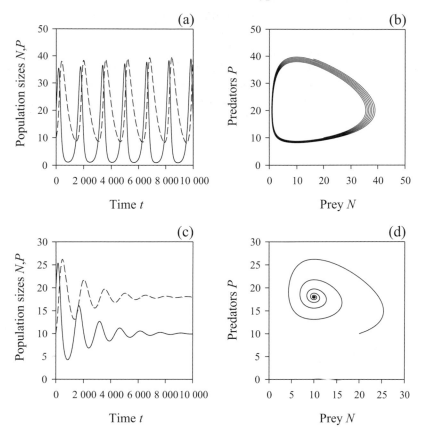

Figure 3.1 Population trajectories obtained from the Lotka–Volterra model: left
panels show prey and predator population sizes N (solid lines) and P (dashed lines) as
functions of time, and right panels show phase diagrams in which P is plotted versus
N over time. The (arbitrary) time unit is days, and the model parameters are: $r = 1/d$,
$\beta = 0.05/d$, $\gamma = 0.02$, $\delta = 0.2/d$, and $\alpha = 0$ (upper panels) and $\alpha = 0.01/d$ (lower
panels). The initial population sizes are $N(t = 0) = 20$ and $P(t = 0) = 10$.

recover and eventually assume a large value again – and the cycle
starts anew.

An example of an economic feedback loop can be found in Armsworth et al. (2006) who analyse feedbacks in the demand and supply of
land. If a conservation agency buys land to conserve it for biodiversity
protection it increases the scarcity of land. Due to the implied rise in
land prices, extensive agriculture that usually delivers ecological benefits
as well becomes less profitable compared with other uses such as hous-
ing. As a consequence, extensive agricultural land is developed into

settlements — an indirect land-use change as in Lapola et al. (2010), presented in Section 3.2. Altogether, conserving land has two effects: the positive effect that biodiversity is protected in the reserves, and the negative effect that biodiversity on the formerly extensive agricultural land is lost. As the authors show, under certain circumstances, especially if the extensively used agricultural land harbours a high level of biodiversity, the conservation of land may on net reduce the overall level of biodiversity.

Part II
Ecological Modelling

4 · *Homogenous Deterministic Population Models*

The present chapter deals with non-spatial ecological population models that do not include any spatial structure and in which all individuals of the population have identical properties. This means, among others, that the location of the individuals and their proximity to other individuals are not considered. Moreover, all individuals have the same age, sex, fitness and so forth. Next to the assumption of homogeneity, stochasticity or randomness play no role.

4.1 Unlimited Population Growth

This section considers the simplest homogenous deterministic ecological population model. The model focuses on the two main processes in a population: reproduction and death. Let $B(t,\Delta t)$ and $D(t,\Delta t)$ denote the number of birth and death events in the population during some time interval Δt between time t and time $t + \Delta t$. If the number of individuals in the population at time t is denoted as $N(t)$, the population size after the time interval Δt will have increased by $B(t)$ and decreased by $D(t)$ and equals:

$$N(t + \Delta t) = N(t) + B(t, \Delta t) - D(t, \Delta t). \tag{4.1}$$

The number of birth events $B(t,\Delta t)$ is the product of three factors: the per capita birth rate b, that is, the number of offspring an individual produces per time unit; the current number of individuals $N(t)$; and the length of the time interval Δt:

$$B(t, \Delta t) = bN(t)\Delta t. \tag{4.2}$$

If, for example, $b = 0.1/\text{month}$, $N(t) = 100$ and $\Delta t = 1$ year, the number of birth events during time interval Δt equals 120. In an analogous manner, the number of death events is the product of the per capita death rate d, that is, the proportion of individuals dying per time unit; the current number of individuals $N(t)$; and the length of the time interval Δt:

$$D(t, \Delta t) = dN(t)\Delta t. \tag{4.3}$$

The population growth equation, Eq. (4.1), then can be written as

$$N(t + \Delta t) = N(t) + B(t, \Delta t) - D(t, \Delta t) \atop = (1 + r\Delta t)N(t)$$ (4.4)

where

$$r = b - d$$ (4.5)

is the per capita population growth rate, that is, the net change per individual per time unit. Rearranging Eq. (4.4) yields

$$\frac{N(t + \Delta t)}{N(t)} = (1 + r\Delta t)$$ (4.6)

which means that during the time interval Δt the population size multiplies by a factor of $1 + r\Delta t$. If $r > 0$ the population size increases exponentially; if $r < 0$ it declines exponentially; and for $r = 1$ it remains constant. A value of $r\Delta t = +1$, for example, implies that the population size doubles during Δt, while for $r\Delta t = -0.5$ it is halved. Time-discrete models in which the population size changes in discrete steps from $N(t)$ to $N(t + \Delta t)$ during discrete time intervals Δt are often used when reproduction and death are synchronised among the individuals of the population, like in many insect or bird species in Central Europe that reproduce only during a short period in summer.

Other species such as humans can reproduce year round and die with a probability that is rather independent of the time of the year. For these species, time-continuous models in which the population size changes continuously are more appropriate. To derive a time-continuous model from Eq. (4.4), one can rearrange to obtain

$$\frac{N(t + \Delta t) - N(t)}{\Delta t} = r$$ (4.7)

which states that the rate of change in the population size, that is, the change in population size between times t and $t + \Delta t$ divided by the length of the time interval Δt, is given by the growth rate r. If Δt is decreased to infinitesimally small values, Eq. (4.7) becomes

$$\frac{dN(t)}{dt} = rN(t),$$ (4.8)

where $dN(t)/dt$ is the derivative of N with respect to time t (not to be confused with the death rate d in Eq. (4.5)). The solution of Eq. (4.8) is

$$N(t) = N(t_0)e^{r(t - t_0)}$$ (4.9)

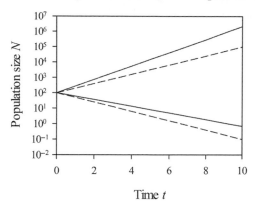

Figure 4.1 Population size N (plotted in logarithmic scale) as a function of time t (in years) for growth rates $r = 1$/year (increasing trajectories) and $r = -0.5$/year (decreasing trajectories). The initial population size is $N(t = 0) = 100$. The solid lines represent time-continuous growth after Eq. (4.9) and the dashed lines represent time-discrete growth after Eq. (4.6).

which represents exponential growth from some initial population size $N(t_0)$ to population size $N(t)$ at time t. Similar to Eq. (4.6), the population grows for $r > 0$, declines for $r < 0$ and remains constant for $r = 0$. For comparison, the population trajectories produced by the two growth models of Eqs (4.6) and (4.8) are shown in Fig. 4.1.

4.2 Limited Population Growth

The previous models of population growth have a severe deficit. Populations either go extinct ($r < 0$) or explode to infinitely large sizes ($r > 0$), and only in the unlikely case in which r is exactly zero do they stay at finite values. A major reason for this unrealistic behaviour is that the models of Eqs (4.6) and (4.8) ignore that population growth r generally depends on the population size. This growth limitation is due to the competition between individuals for resources such as food, water, space, light and so forth. Overcrowding in large populations implies that some or all individuals get fewer resources than required to realise their maximum reproductive success and lifespan.

The simplest assumption to model competition among the individuals in a population is to assume a fixed resource or carrying capacity K. If $N(t)$ is the population size at time t then $K/N(t)$ is each individual's current (average) share of the resource. If that share gets smaller, so does

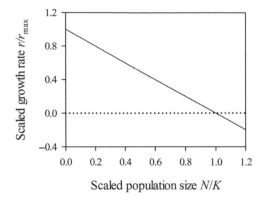

Figure 4.2 Example for the dependence of the current per capita growth rate $r(t)$ on the current population size $N(t)$. The growth rate r is scaled in units of its maximum value r_{max} and N is scaled in units of the carrying capacity K.

the per capita growth rate r. The simplest model for this relationship is the logistic growth model which assumes a linear dependence of r on N, so that r has a maximum value r_{max} (termed the intrinsic growth rate) for $N = 0$ and declines to zero when N approaches K:

$$r(t) = r_{max}\left(1 - \frac{N(t)}{K}\right). \tag{4.10}$$

A graphical example is shown in Fig. 4.2, which also includes the possibility of population sizes above the carrying capacity that imply negative population growth, $r < 0$. Since the growth rate r depends on the density of individuals, N/K, the population is said to be subject to density-dependence.

Inserting Eq. (4.10) into Eq. (4.8) yields a differential equation for logistic population growth:

$$\frac{dN(t)}{dt} = rN(t) = r_{max}N(t)\left(1 - \frac{N(t)}{K}\right) \tag{4.11}$$

which can be solved to

$$N(t) = \frac{K}{1 + (K/N(t_0) - 1)\exp\left\{-r_{max}K\cdot(t - t_0)\right\}}. \tag{4.12}$$

Some numerical examples are shown in Fig. 4.3. Starting from small population sizes $N \ll K$, the population grows exponentially as

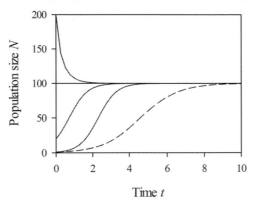

Figure 4.3 Population size N as a function of time t after Eq. (4.12). Solid lines: different initial population sizes $N(t_0)$ and intrinsic growth rate $r_{max} = 0.02$; dashed line: $r_{max} = 0.01$. Carrying capacity $K = 100$.

in Fig. 4.1 until, with increasing population size, the growth rate declines and the population size asymptotically approaches the carrying capacity K.

The population model of Eq. (4.11) can also be formulated in discrete time with time steps Δt. For this the time derivative $dN(t)/dt$ is replaced by the fraction $[N(t + \Delta t) - N(t)]/\Delta t$ (the reverse of the reformulation from Eq. (4.7) to Eq. (4.8)), and Eq. (4.11) becomes

$$\frac{N(t + \Delta t) - N(t)}{\Delta t} = rN(t) = r_{max}N(t)\left(1 - \frac{N(t)}{K}\right) \qquad (4.13)$$

which can be rearranged to

$$N(t + \Delta t) = N(t)\left(1 + r_{max}\Delta t\left(1 - \frac{N(t)}{K}\right)\right). \qquad (4.14)$$

If the product $r_{max}\Delta t$ is small, which implies that the population size changes only marginally during the time interval Δt, the population trajectory (Fig. 4.4a) has the same shape as that in Fig. 4.3 where, according to the definition of a time-continuous model, the population changes only marginally during each time interval. However, if $r_{max}\Delta t$ is increased, oscillations appear. The so-called periodicity of these oscillations starts with 2 (Fig. 4.4c) for comparatively small values of $r_{max}\Delta t$ and increases with increasing $r_{max}\Delta t$ (Fig. 4.4e).

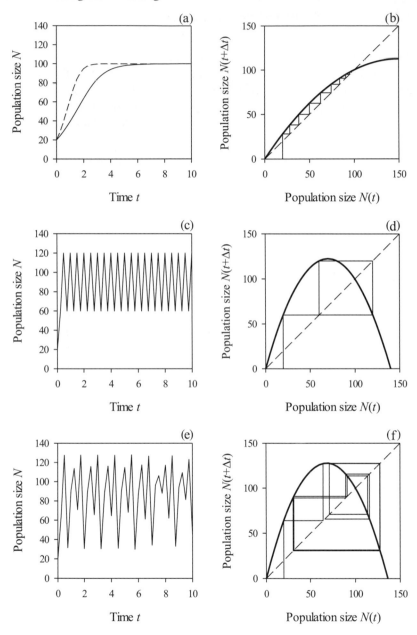

Figure 4.4 Population size N as a function of time t after Eq. (4.14) (left panels) and corresponding phase diagrams (right panels). Upper panels: $r_{max} = 1$ (dashed line in panel a: $r_{max} = 2$); middle panels: $r_{max} = 10$; lower panels: $r_{max} = 11$. Carrying capacity $K = 100$, initial population size $N(t_0) = 20$, length of time step $\Delta t = 0.25$.

4.3 Chaos and Scramble Competition

The mathematical reason for these oscillations is the shape of the map $N(t) \rightarrow N(t + \Delta t)$ that shows which population size $N(t + \Delta t)$ is obtained after Δt if the current population size is $N(t)$ (Fig. 4.4b,d,f). Consider Fig. 4.4d with an initial population size $N(t_0) = 20$. The following value $N(t_0 + \Delta t) = 60$ can be read from the graph (bold line). The value $N(t_0 + 2\Delta t)$ following that is obtained by moving from the point $(N(t), N(t + \Delta t)) = (20, 60)$ rightwards to the diagonal $(60, 60)$, and from there down to the $N(t)$-axis. Since the diagonal is defined by the equality $N(t + \Delta t) = N(t)$, the end point of this move is the point $(60, 0)$, that is, $N(t) = 60$. Now the value $N(t + \Delta t)$ with $t = t_0 + \Delta t$ can be read from the graph again by moving upwards to the bold line, ending at point $(60, 120)$, so $N(t_0 + 2\Delta t) = 120$. The next value $N(t_0 + 3\Delta t) = 60$ is found in the same manner, as well as all later values, cycling anticlockwise along the square in the figure.

As can be seen in Fig. 4.4f, the movement can be quite erratic and may not converge into a simple trajectory. Such dynamics are called chaotic (May 1976; Alligood et al. 2000), since they are unpredictable at least in the long run. Simulations of Eq. (4.14) with the parameters of Fig. 4.4e,f show that even tiny deviations in the initial population size lead to large differences in the predicted population sizes already after a few time steps.

Why do large values of $r_{max}\Delta t$ lead to oscillations or even chaotic dynamics? A necessary condition for these types of dynamics is the hump in the map $N(t) \rightarrow N(t + \Delta t)$ (bold lines in Fig. 4.4d,f). Although the map in Fig. 4.4b exhibits a hump, too, its maximum occurs only at large values of $N(t)$ that are never assumed in the course of the dynamics. Instead, within the range of possible population sizes, the map increases monotonously, which implies a smooth convergence of the population size towards the intersection of the map $N(t) \rightarrow N(t + \Delta t)$ and the diagonal.

This observation has important ecological implications. Ecologists distinguish between two types of intraspecific competition: contest and scramble (Begon et al. 1990, ch. 6.5). In contest competition – as the name indicates – a few winning individuals get all the available resource (carrying capacity K) and survive and reproduce with high probability, while all the others starve and die with high probability. Since an 'excess' of individuals does not affect the survival and reproduction of the winners, a larger current population size $N(t)$ never harms it and an increase in $N(t)$ either increases $N(t + \Delta t)$ or does not affect it, but it never leads to a decrease in $N(t + \Delta t)$ (bold line in Fig. 4.4b). As a

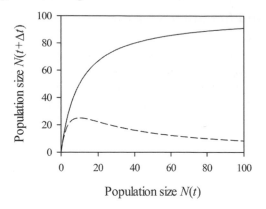

Figure 4.5 Two numerical examples of the Hassell model (Eq. 4.15). Solid line: $c = 1$; dashed line: $c = 2$. Parameters $k_1 = 10$ and $k_2 = 0.1$.

consequence, if a population is limited by contest competition the population size smoothly approaches the carrying capacity (Fig. 4.4a,b).

In scramble competition, in contrast, there are no clear winners and losers but all individuals receive some share of the resource. If the current number of individuals is low, that share is high and most of the individuals survive and reproduce. Similar to the case of contest competition, larger $N(t)$ implies larger $N(t + \Delta t)$ (left side of the hump-shaped curve in Fig. 4.4d). However, if the current population size is large, the share of the resource does not suffice for *any* individual (or only for very few individuals) and the number of surviving and reproducing individuals is small, and the larger the current population size $N(t)$ the smaller the next population size $N(t + \Delta t)$ (right side of the hump-shaped curve in Fig. 4.4d). Altogether, a comparatively small population size $N(t)$ is followed by a comparatively large population size $N(t + \Delta t)$, which is followed by a comparatively small size $N(t + 2\Delta t)$, and so on (Fig. 4.4c,d).

Various functions have been constructed to describe contest and scramble competition (Hassell 1975) and the transition between the two, such as the Hassell model (Geritz and Kisdi 2004):

$$N(t + \Delta t) = N(t)\frac{k_1}{(1 + k_2 N(t))^c}. \tag{4.15}$$

Figure 4.5 shows two numerical examples. Whether a population is limited by contest or scramble competition can have an important influence on the survival of the population and its dependence on land use. Johst et al. (2006) modelled the impact of mowing on the survival of two butterfly species, *Maculinea nausithous* and *M. teleius*, which are regarded as being limited by contest and scramble competition, respectively. While survival of *M. nausithous* was maximised by mowing regimes that impose only little mortality on eggs and larvae, *M. teleius* survived best under mowing regimes associated with some moderate mortality. The reason is that this moderate mortality prevents the scramble competitor *M. teleius* from reaching too large population sizes which would be followed by severe population crashes and an increased population extinction risk.

Recent literature related to Chapter 4 includes Wakano et al. (2009), M'Gonigle and Greenspoon (2014) and Barraquand et al. (2017).

5 · *Homogenous Stochastic Population Models*

5.1 Stochasticity in Population Dynamics

Is the world stochastic or deterministic? Is everything that happens predetermined or is there room for randomness? Answering or even addressing this difficult question is clearly beyond the scope of the present chapter. A simpler question is: is it sensible to consider and model some processes in the real world as if they were stochastic – and the answer is 'yes'.

To understand the reason, consider one of the oldest random number generators humankind has invented: the die. The rolling of a die starts with the die being located at a particular location with a particular orientation in the player's hand with the hand having a particular location and orientation relative to the table; this initial condition is followed by a particular movement of the hand, and the die rolls over the table and ends its movement at a particular location and orientation on the table. In principle, the initial condition of the process can be perfectly measured, and the process follows simple and well-known deterministic physical laws. The problem is that in practice we do not know the initial condition with perfect precision, nor do we know exactly the movement of the hand, nor the boundary conditions of the process, such as the shape of and mass distribution within the die, the physical properties of the table's surface and more. As a consequence, although the process of rolling a die is entirely deterministic and its outcome perfectly predetermined, in practice it appears random, and so dice work reasonably well as random number generators.

A more theoretical but mathematically even stricter example of a deterministic process that only appears random is the chaotic dynamics discussed in Section 4.3. Equation (4.14) clearly does not contain any random element, but the dynamics it produces in Fig. 4.4f appear random. The same can be said about many processes in ecosystems. These systems are usually so complex, affected by so many processes interacting in a complex manner, that even if these processes are entirely deterministic, they do appear, and can (to quite some extent) be treated, as stochastic processes.

Referring to the homogenous population model in Chapter 4, stochasticity can be effective on two levels. First, note that the per capita death rate d used in Eq. (4.3) is an average over all individuals of the population. However, due to differences among the individuals, such as age, health, nutritional state, spatial location and others, some individuals may have a higher and others may have a lower likelihood of dying. Since we do not know all these details about the individuals, we consider the death rate d as stochastic and define it as an individual's probability of dying per time unit. A value of $d = 0.1$/month, for example, means that the individual dies within a month with probability 0.1. In the same way, the birth rate b (Eq. 4.2) is defined as the probability of producing an offspring per time unit. Since this type of stochasticity is due to demographic differences among individuals, it is termed *demographic stochasticity*.

In addition to this demographic stochasticity, the magnitudes of the stochastic rates b and d themselves may vary randomly. Reasons for this include unpredictable variation in food resources, weather and other environmental factors that affect all individuals in the population in the same correlated manner. Therefore, this type of stochasticity is termed *environmental stochasticity*.

5.2 Probability Distributions and Random Numbers

To include stochasticity in the deterministic population model of Chapter 4, some basic knowledge is required about relevant probability distributions and about how to draw random numbers from these distributions (Epps 2015). The most well-known probability distribution is probably the normal distribution, which is defined by

$$\varphi(x) = \frac{1}{\sigma\sqrt{2\pi}} \exp\left\{ -\frac{(x - \mu)^2}{2\sigma^2} \right\}. \tag{5.1}$$

Quantity $\varphi(x)$ is a probability density (cf. Fig. 5.1), so that the integral $\int \varphi(x)dx$ between two bounds x_1 and x_2 (i.e. the area under the curve between x_1 and x_2) is the probability of observing an x-value between x_1 and x_2. The quantities μ and σ are the mean and the standard deviation of the distribution, respectively.

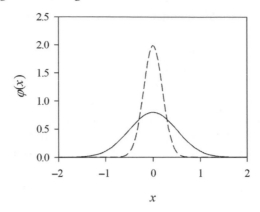

Figure 5.1 Probability density $\varphi(x)$ of the normal distribution, Eq. (5.1).
Mean $\mu = 0$ and standard deviation $\sigma = 0.5$ (solid line) and $\sigma = 0.2$ (dashed line).

The popularity of the normal distribution for the modelling of random environmental variables probably stems from the fact that environmental variables themselves are often determined by several random factors. According to the so-called central limit theorem, under certain conditions the sum of many random numbers is normally distributed, and so it is a reasonable assumption for many random environmental variables to be normally distributed.

To draw a random number from a normal distribution, one first has to build the cumulative normal distribution

$$\Phi(x) = \frac{1}{\sigma\sqrt{2\pi}} \int\limits_{-\infty}^{x} \exp\left\{ -\frac{(x' - \mu)^2}{2\sigma^2} \right\} dx' \tag{5.2}$$

which represents the probability of observing a value equal to or below x (Fig. 5.2).

Having constructed the cumulative normal distribution, a random number has to be drawn from the *uniform* distribution in which all values are equally likely, with lower and upper bounds of 0 and 1, respectively. The result may be, for example, 0.750. In the example of Fig. 5.2 this value is identified with $\Phi(x)$ of the cumulative normal distribution that corresponds to $x = 0.325$ (Fig. 5.2). This value is the sought random number (for an explanation, see discussion of Fig. 5.3b).

The second probability distribution of major interest is the binomial distribution. It is applied if one wants to know how often a particular event (e.g. a thrown coin landing heads up) that occurs with a probability p (0.5 in the case of the coin) is observed in a certain number of trials (throws

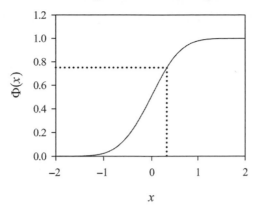

Figure 5.2 Cumulative normal distribution $\Phi(x)$, Eq. (5.2). Mean and standard deviation are $\mu = 0$ and $\sigma = 0.5$, respectively. The dotted line represents the functional relationship $\Phi(x) = 0.325 = 0.750$.

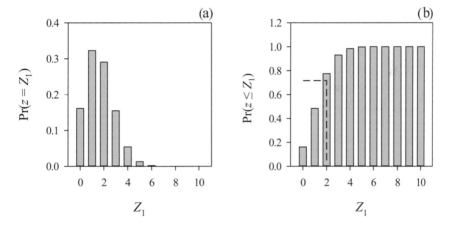

Figure 5.3 Binomial distribution and cumulative binomial distribution. The bars show the probability of observing the event of interest Z_1 times (left panel) and Z_1 times or fewer (right panel). The number of trials is $Z = 10$ and the probability of observing the event is $p = 1/6$. For the dashed line, see text.

of the coin). If an event occurs with probability p then the probability of observing it Z_1 times out of Z trials is binomially distributed with

$$\Pr(Z_1, p, Z) = \mathrm{Bin}(Z_1, p, Z) = \frac{Z!}{Z_1!(Z - Z_1)!} p^{Z_1}(1 - p)^{Z - Z_1}. \quad (5.3)$$

An example is shown in Fig. 5.3a. If a die is rolled ten times and the probability of getting a '6' is $p = 1/6$ then the probability of getting the '6' twice is $\mathrm{Bin}(2, 1/6, 10) \approx 0.28$. The mean of the binomial distribution equals pZ, so if the die is rolled ten times the average number of throws with a '6' is $10/6 \approx 1.67$, and the variance is $Zp(1-p) = 10 \times 1/6 \times 5/6 \approx 1.39$.

To draw a random number from the binomial distribution, similar to the drawing of a random number from the normal distribution (Figs 5.1 and 5.2), the cumulative binomial distribution needs to be built which represents the probability of observing the event Z_1 times or fewer (Fig. 5.3b). Once again, a uniformly distributed random number (u) is drawn from the interval [0,1]. According to Fig. 5.3b, the probability of observing $Z_1 = 0$ is 0.16, so if u happens to be equal to or smaller than 0.16 we conclude that $Z_1 = 0$. Analogously, according to Fig. 5.3b the probability of observing $Z_1 \leq 1$ is 0.48 and if u happens to be equal to or smaller than 0.48 we conclude that Z_1 is either zero or one. If at the same time $u > 0.16$, we know from the former consideration that $Z_1 > 0$, so altogether the only possible conclusion is $Z_1 = 1$, and a value of u in the interval (0.16, 0.48] implies $Z_1 = 1$. This consideration can be generalised to any value of u so that if $u \in (\Pr(z \leq a - 1), \Pr(z \leq a)]$ with $a \in [0,1]$ the sought random number from the binomial distribution is $Z_1 = a$, or graphically: identify the bar in the cumulative distribution that is as short as possible but above the random number u. If, for instance, u happens to be 0.72 then for the distribution in Fig. 5.3b the sought random number is $Z_2 = 2$. The sampling procedure from the normal distribution described previously can be justified in an analogous manner.

If the number of trials, Z, becomes very large, the numerical calculation of the factorials in Eq. (5.3) may become tedious or even impossible. In these cases the factorial might be approximated by closed formulas. Or, even more simply, one can apply the Moivre–Laplace Theorem, which states that for sufficiently large Z the binomial distribution can be approximated by a normal distribution with the mean and variance of pZ and $Zp(1-p)$, respectively, as provided in the discussion of Fig. 5.3a.

The third and last relevant probability distribution is the Poisson distribution, which has the following application. If an event occurs on average λ times within some (time) interval then the probability of the event occurring L times during the interval is

$$\Pr(L, \lambda) = \text{Pois}(L, \lambda) = \frac{\lambda^L}{L!} e^{-\lambda}. \tag{5.4}$$

An example is shown in Fig. 5.4. Mean and variance of the Poisson distribution are equal and given by λ. A random number can be drawn

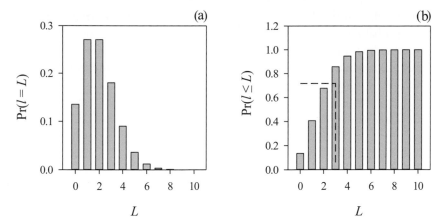

Figure 5.4 Poisson distribution and cumulative Poisson distribution. The bars show the probability of observing the event of interest L times (a) and L times or fewer (b) in the interval. On average the event occurs $\lambda = 2$ times in the interval. For the dashed line, see the explanation of Fig. 5.3b.

from the Poisson distribution by forming its cumulative distribution (Fig. 5.4b) and proceeding as described previously for the binomial distribution.

5.3 Simulating Stochastic Population Dynamics and Population Viability Analysis

With the methods of Section 5.2 at hand, the stochastic dynamics of a single homogenous population can be modelled and simulated as follows. The starting point is the model described by Eqs (4.1) and (4.2). Consider a population at some time t with size $N(t)$ and the first process of the population dynamics: the birth events. With b being the number of offspring produced per individual per time unit, the average number of offspring produced during time interval Δt is $\lambda = bN(t)\Delta t$. The number of offspring produced in a population is usually modelled via the Poisson distribution, and so the probability of L offspring being produced is (cf. Eq. 5.4)

$$\Pr(L, bN(t)\Delta t) = \mathrm{Pois}(L, bN(t)\Delta t) = \frac{(bN(t)\Delta t)^L}{L!} e^{-bN(t)\Delta t}. \qquad (5.5)$$

The actual number of offspring, L, is then sampled from Eq. (5.5) as demonstrated in Section 5.2. The second process, the deaths of individuals, is modelled using the binomial distribution. With d being the rate (probability per time unit) of an individual dying, the probability of an individual surviving the time interval Δt is $1 - d\Delta t$, and the probability of Z_1 individuals surviving from a population of $N(t)$ individual is (cf. Eq. 5.3)

$$
\begin{aligned}
\Pr(Z_1, 1 - d\Delta t, N(t)) &= \text{Bin}(Z_1, 1 - d\Delta t, N(t)) \\
&= \frac{N(t)!}{Z_1!(N(t) - Z_1)!} (1 - d\Delta t)^{Z_1} (d\Delta t)^{N(t) - Z_1}.
\end{aligned}
$$

$$(5.6)$$

Again, the actual number of survivors, Z_1, is sampled from Eq. (5.6) as previously demonstrated. After the two processes of reproduction and survival, the population size after time interval Δt is (Eq. 4.1)

$$
N(t + \Delta t) = N(t) + L + Z_1. \tag{5.7}
$$

From this population size $N(t + \Delta t)$, the population size $N(t + 2\Delta t)$ after another time interval Δt is calculated using Eqs (5.5) and (5.6) as described previouslyby replacing $N(t)$ in Eqs (5.5) and (5.6) with $N(t + \Delta t)$. This process can be repeated to simulate the population dynamics for an arbitrary number of time intervals.

Two notes might be added here. First, to keep the consistency with Eq. (4.1), it was assumed in this simulation that the L offspring survive with certainty till the end of the time interval Δt, or if in contrast offspring could die within the time interval, these deaths would be implicitly included in the birth rate b. Alternatively, one could subject both adults and offspring to the mortality d; this, however, would be a different model than the one reprcsented by Eq. (4.1).

Second, density–dependence was, for simplicity, not considered in this simulation. However, it could be included in a straightforward manner, for example by replacing b with $b(1 - N/K)$ with some carrying capacity K as in Eq. (4.10).

Typical outcomes of the simulation with Eqs (5.5) and (5.6) are shown in Fig. 5.5. The trajectories differ considerably and reflect the stochasticity in the birth and death processes. This implies that no single trajectory can be used to make any prediction on the future of the population, nor can it be said whether the population will survive for some time or not. Instead, the future of the population can only be described properly in a statistical manner.

A first approach for this is to calculate for each time step the mean and standard deviation of the population size, taking into account a

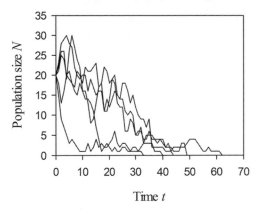

Figure 5.5 Five random population trajectories $N(t)$ starting from an initial population size of $N(0) = 20$ with per capita birth and death rates of $b = 0.35$ per time step and $d = 0.4$ per time step, respectively.

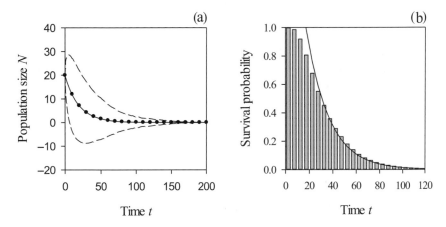

Figure 5.6 Statistics of the population dynamics based on 10,000 simulation runs with the parameters of Fig. 5.5. Panel a: mean population size (solid line) and mean population size plus/minus two standard deviations (dashed lines); the dots show the deterministic dynamics of the model (see text). Panel b: Probability of surviving (at least) till time t for the simulation runs of panel a, and fit (solid line) according to Eq. (5.14) (see text).

sufficiently large number of simulation replicates. In Fig. 5.6a one can see that the simulated mean population size agrees very well with the deterministic dynamics after Eq. (4.1), where at every time step the population size multiplies by a factor of $1 + b - d = 0.95$. However, the variation in the population size is very large and a lower bound, defined here as the mean minus two standard deviations, is negative most of the time. In such a situation it is doubtful whether the mean population size is a sensible measure of the state of the population, and whether the development of that mean allows for any sensible assessment of the viability of the population.

Assessing the viability of populations, termed population viability analysis, is a prominent research field in conservation biology (Burgman et al. 1993; Akçakaya et al. 1999; Coulson et al. 2001; Beissinger and McCullough 2002). The first step towards a meaningful population viability analysis is to simulate the population dynamics I times and record in each simulation run i the time step τ_i at which the population went extinct, that is, the time step at which it assumed the value $N = 0$. A good measure of population viability, then, is the mean population life time

$$T = \frac{1}{I} \sum_{i=1}^{I} \tau_i. \tag{5.8}$$

For the above numerical example, the mean population life time can be determined as $T = 37.3$. The fact that the viability of the population can be expressed by a single number makes this viability measure a very handy one. The disadvantage is that it does not tell us much about the actual extinction risk of the population. This is better reflected in the probability distribution of the extinction times, $p(t)$, where $p(t)\Delta t$ is the proportion of extinction events observed in the time interval $[t, t + \Delta t]$. The integral

$$P(t) = 1 - \int_0^t p(t')dt' \tag{5.9}$$

yields the probability of the population surviving till time t (Grimm and Wissel 2004). Figure 5.6b shows this survival probability for the population model of Fig. 5.5. One can see that in the first few time steps the survival probability is (close to) one, so no or very few extinction events take place, but afterwards the survival probability declines exponentially with time t,

$$P(t) = e^{-t/T}, \tag{5.10}$$

at a rate $1/T$ which is the inverse of the mean population life time T (Section 5.4). The shape of the distribution shown in Fig. 5.6b is typical and has been observed in a large number of population viability analyses. In fact, it can be justified mathematically (Section 5.4).

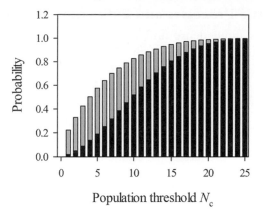

Figure 5.7 Quasiextinction risk: probability of falling below a population threshold N_c within 10 time steps (black bars) and 20 time steps (black plus grey bars). Model parameters as in Fig. 5.5.

Equation (5.10) is well suited to set conservation targets for endangered species (cf. Burgman et al. 1993). For instance, one can specify a time horizon t_H and demand that the population survives for this amount of time at least with a certain minimum probability P_{min}. Or alternatively, one can specify a minimum value for the survival probability and demand that this is sustained over a certain time horizon.

In the practice of biodiversity conservation, the probability of complete extinction, $N = 0$, is sometimes of less interest, because when the population has become close to that state it is usually way too late for any intervention to secure the species. To consider that many species can be counted as practically extinct already at larger population sizes, $N > 0$, the concept of quasiextinction risk has been introduced (Ferson and Burgman 1995), which represents the risk of the population falling below some critical level N_c during some time horizon t_H. Two examples for the population model of Fig. 5.5 are shown in Fig. 5.7. The quasiextinction risk curves have a concave or sigmoid shape, so that the probability of falling below a very small population size is small but at some level increases rapidly with increasing threshold N_c until it saturates at a value of 1.

So far the focus has been on demographic stochasticity. Environmental stochasticity can be included in a straightforward manner, for example by assuming that the demographic rates b and d are not constant but drawn in each time step from normal distributions with certain means m_b and m_d and standard deviations σ_b and σ_d.

5.4 Stochastic Processes and Extinction Risk

This section is quite dense and mathematically demanding. It is not essential for understanding the other content of this book, but for the interested reader it presents some mathematical foundations of the stochastic population model analyses of Section 5.3.

Besides simulation, stochastic processes can also be described and analysed through stochastic differential equations (Goel and Richter-Dyn 1974; Nisbet and Gurney 2003). Similar to Eq. (4.8), these equations describe how the rate at which the focal system (the population size in the example of Eq. (4.8)) changes over time depends on the system's current state. However, while deterministic differential equations such as Eq. (4.8) describe how the state variable(s) of the system change in time, stochastic differential equations describe how the *probability(s)* of the state variable(s) assuming particular values change(s) with time. One of the most well-known stochastic differential equations is the so-called Master equation for a system that has discrete states and changes continuously in time:

$$\frac{dP_n(t)}{dt} = \sum_m a_{mn} P_m(t). \tag{5.11}$$

Here $P_n(t)$ is the probability of observing the system in state n, and a_{mn} represents the elements of a transition matrix **A** that describe the transition of the system from some state m to state n. If n is considered the current size of a stochastically changing population and if one assumes that the birth and death events in this population are uncorrelated, then from some population size n only changes to population sizes $n + 1$ and $n - 1$ are possible, and Eq. (5.11) becomes

$$\frac{dP_n(t)}{dt} = b_{n-1}P_{n-1}(t) + d_{n+1}P_{n+1}(t) - (b_n + d_n)P_n(t). \tag{5.12}$$

Here the first summand on the right-hand side describes the probabilistic change from population size $n - 1$ to n (which adds to the probability of observing population size n), the second summand describes the change form $n + 1$ to n, and the third summand describes the change from n to $n + 1$ and $n - 1$, respectively. The dynamics generated by a Master equation such as Eq. (5.12) are called a Markov process and are mainly characterised by the facts that transitions between the system states are random and that the transitions between two consecutive time steps only depend on the system state in the former time step (i.e. the process has no memory).

The solution of Eq. (5.12) can be shown to be

$$P_n(t) = \sum_i u_{ni} c_i e^{-\omega_i t},$$ (5.13)

where u_{ni} is the n-th component of the normalised right-hand eigenvector of matrix $\mathbf{A}$ corresponding to the i-th eigenvalue ω_i, and c_i is the inner product of the corresponding left-hand eigenvector with the initial condition $P_n(t = 0)$ (Grimm and Wissel 2004). In many cases the first eigenvalue ω_1 is much smaller than all the other eigenvalues and with some further arguments the probability $P_{n>0}$ of observing at least one individual in the population becomes (after a short transition time)

$$P_{n>0}(t) = c_1 e^{-\omega_1 t}$$ (5.14)

(Grimm and Wissel 2004). As Eq. (5.14) shows, the probability of observing one or more individuals declines exponentially with time. The minimum of c_1 and 1 is approximately the probability of the population reaching the so-called quasistationary state in which the population exhibits typical fluctuations around its mean. Quantity ω_1 is the rate at which the probability of observing a non-zero population size declines with time. Equivalently, it can be identified with the inverse of the expected life time T of this state, that is of the mean population life time introduced in the Section 5.3.

Plotting $P_{n>0}$ in logarithmic scale as a function of t leads to a straight line with slope $-\omega_1 = -1/T$ and offset $\ln(c_1)$. For the model of Fig. 5.5, this plot is shown in Fig. 5.8. Linear regression delivers the fitted survival probability

$$\ln P(t) \equiv \ln P_{n>0}(t) \approx 0.894 - \frac{t}{19.3},$$ (5.15)

indicating a mean population life time of 19.3. Taking the exponential of this equation yields the solid line in Fig. 5.6b.

The mean population life time of 19.3 identified in Eq. (5.15) substantially differs from the average extinction time over the 10,000 simulation runs which had been calculated in Section 5.3 to 37.3. The reason for this difference is that the value of 19.3 is valid for a population that is initially in its quasistationary state, while the simulations of Section 5.3 do not start with the population residing in this quasistationary state. Instead, the initial population size $N(0) = 20$ of the simulation was 'too large', so it took $37.3 - 19.3 = 18.0$ time steps for the population to reach the quasistationary state from which it took on average those 19.3 time steps to go extinct (for details, see Grimm and Wissel 2004).

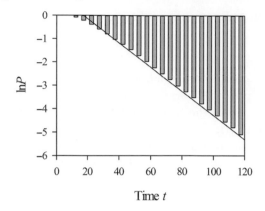

Figure 5.8 Probability of the population surviving (at least) till time t (cf. Fig. 5.6b) on logarithmic scale. The solid line represents a linear fit with offset 0.894 and slope $-0.0518 = -1/19.3$.

The Master equation (Eq. 5.12) can be used to analytically calculate the mean life time of the population (Wissel and Zaschke 1994). The authors equate the difference between the birth and death rates with logistic growth,

$$b_n - d_n = m\left(1 - \frac{n}{K}\right),\qquad(5.16)$$

where r is the intrinsic growth rate and K the carrying capacity (cf. Section 4.2), and model the sum of the birth and death rates via

$$b_n + d_n = \eta n + \sigma^2 n^2,\qquad(5.17)$$

where η represents the demographic stochasticity caused by the randomness of the birth and death events, and σ^2/r is the squared coefficient of variation of the intrinsic population growth rate, representing environmental stochasticity. Inserting Eqs (5.16) and (5.17) into Eq. (5.12) and calculating the leading eigenvalue ω_1 of the transition matrix **A** (cf. Eq. 5.11) delivers with some algebra

$$T = \frac{1}{\omega_1} \propto \begin{cases} \ln K & \sigma^2/r \ge 2 \\ K^{2r/\sigma^2-1} & \sigma^2/r < 2 \end{cases}\qquad(5.18)$$

(Wissel and Zaschke 1994). In most real-world cases $\sigma^2/r < 2$ should be valid, so that the mean population life time increases algebraically with increasing carrying capacity K, and the exponent $2r/\sigma^2 - 1$ is inversely related to the strength of the environmental stochasticity.

5.5 The Risk Model of the IUCN

As demonstrated previously, stochastic population models can be used to assess the viability of populations. Population viability may be measured through different quantities, such as mean and variation of population size and the risk of becoming extinct within a certain time period. Such risk assessments are an important fundament of the classification of the conservation status of species by the International Union for Conservation of Nature (IUCN) which feeds into IUCN's well-known Red List of Threatened Species. In this list, each species for which adequate data is available is assigned to one of seven categories (Rodrigues et al. 2006). These are, in increasing order of extinction risk: least concern, near threatened, vulnerable, endangered, critically endangered, extinct in the wild and extinct.

A number of criteria are used by IUCN to classify a species (Rodrigues et al. 2006):

1. reduction of population size (per cent during time unit),
2. range of occurrence (km^2, which may include occupied and empty areas),
3. range of occupancy (km^2),
4. population size (number of individuals) and
5. extinction risk (per cent during time unit).

This list is very simplified and contains only the main issues from the full IUCN list, which also considers, among others, the degree of spatial fragmentation of the population and whether the causes of population reduction are reversible or not.

Each criterion includes specifications of levels beyond which a species is classified within a particular conservation status. For instance, if the species declines by at least 30, 50 or 80 per cent within ten years or three generations, it is classified as vulnerable, endangered or critically endangered, respectively, in the criterion 'reduction in population size'. Such an assignment is carried out for each criterion and the most pessimistic classification is used to define the conservation status of the species altogether.

In the context of stochastic modelling and population viability analysis the most interesting criterion is the last one in the list provided here: if a species has an estimated extinction risk of at least 10 per cent in 100 years, 20 per cent in 20 years/five generations or 50 per cent in ten years/three generations, the species is categorised as vulnerable, endangered or critically endangered, respectively, in the criterion 'extinction risk'. With the approximation of Eq. (5.10), these levels correspond to mean population life times of 949, 89.6 and 14.4 years, respectively.

The risk model described by IUCN differs from the mechanistic models featured in this book, because the species' conservation status is not quantitatively deduced from first principles of population ecology. Due to its simplicity (compared with a complex mechanistic model) it can provide only a rough estimate of a species' extinction risk. For instance, some species that occur in high densities may persist quite well in an occurrence range that may be much too small for other species that occur at small densities. Similarly, how strongly a change in a species' occurrence range affects the probability of extinction in the next 50 years very much depends on the species' ecology (e.g. means and variances of birth and death rates, population structure and spatial structure).

On the other hand, IUCN's risk model can utilise a lot of expert knowledge that cannot easily be cast into a mechanistic model. Overall, data deficiencies and time constraints preclude the construction of a mechanistic model for each of the species and all its geographic locations currently contained in the Red List. Another strength of IUCN's model is that it uses very different population–ecological criteria, such as temporal trend (dynamic pattern) and range of occurrence (spatial pattern), whose combined use makes the risk assessment much more reliable than if only a single criterion was employed.

The risk model of the IUCN can be used not only to assess the conservation status of a species but also to guide the selection of conservation areas. For instance, one might prioritise areas that harbour threatened species in significant numbers, contain a significant proportion of the species' ranges and are under threat of becoming lost, for example due to economic development (Eken et al. 2004; Rodrigues et al. 2004). The concept can be extended further from a single species to the risk assessment of whole ecosystems (Rodríguez et al. 2010).

Recent literature related to Chapter 5 includes McElderry et al. (2015), Sharma et al. (2015), Laufenberg et al. (2016), Dennis et al. (2016), Robinson et al. (2017), McGowan et al. (2017) and Fung et al. (2018).

6 · *Spatial Population Models*

The models in Chapters 4 and 5 are spatially homogenous in that the environment inhabited by the population has no spatial structure and the interactions of the individuals with each other and with their environment are independent of the individuals' spatial locations. However, many ecological processes depend on the locations of and distance between entities (Tilman and Kareiva 1997; Jopp et al. 2011). Two main approaches exist for the consideration of spatial structure: patch-based models and grid-based models. In patch-based models landscapes are assumed to consist of habitable land patches in which individuals can survive and reproduce, surrounded by a hostile matrix through which individuals can disperse but in which survival and reproduction is impossible. A habitat usually harbours several individuals which form a local population. Habitat patches may differ in size and quality for the species, but within individual patches conditions are usually assumed to be spatially homogenous.

In grid-based models the landscape is structured as a grid of (usually square, sometimes hexagonal) cells. The cells may have different properties, such as quality for reproduction and survival of individuals. In contrast to the patches in patch-based models, a grid cell with suitable habitat need not necessarily be surrounded by a hostile matrix but may be adjacent to other grid cells with non-zero habitat quality. Consequently, grid-based landscapes usually have a more complex, 'fractal' structure than patch-based landscapes. In addition, while patches in patch-based models are usually large enough to provide space for several individuals, the cells in a grid-based model may provide space for only a single individual such as a tree. Therefore, in patch-based models the biological entity of interest is usually the population while in grid-based models the focus is often on the individual.

6.1 Patch–Based Models: The Metapopulation Concept

One of the most widely used approaches of spatial ecology is the metapopulation concept. Touched on briefly in Section 3.1, this assumes that the landscape consists of habitable patches surrounded by a hostile matrix. Each habitat patch may be occupied by a local population or be empty. The population dynamics within individual patches and the question of whether the local population is currently large or small is often ignored. Instead the focus is on two regional-scale processes: the extinction of local populations and the colonisation of empty patches by local populations (Hanski and Gilpin 1997; Hanski and Gaggiotti 2004).

The simplest metapopulation model is by Levins (1969), as mentioned in Section 3.1. If e is the rate by which local populations go extinct and s the current number of local populations, then es local populations go extinct per time unit. If c is the rate by which a local population colonises a patch, then cs patches are colonised per time unit. Assuming there are altogether S habitat patches in the landscape, then $1 - s/S$ is the proportion of empty patches, and the number of empty patches colonised and turning into occupied patches per time unit is $cs(1 - s/S)$.

The change in the number of local populations, then, is the difference between the number of colonisation events and the number of extinction events and can be described by the differential equation

$$\frac{ds}{dt} = cs\left(1 - \frac{s}{S}\right) - es. \tag{6.1}$$

The steady state in which the number of local populations does not change in time is found by setting the time derivative to zero: $ds/dt = 0$. Solving this equation for the number of local populations, s^*, in the steady state yields

$$s^* = S\max\left[1 - \frac{e}{c}, 0\right] \tag{6.2}$$

which is zero if the extinction rate e exceeds the colonisation rate c, and otherwise increases in a concave manner with increasing 'turnover ratio' c/e. For very large c/e it asymptotically approaches its maximum S (Fig. 6.1). In the context of biodiversity conservation, a necessary condition for a viable metapopulation thus is that c exceeds e.

Levins' motivation in developing this model was to demonstrate the role of regional dynamics for the size and survival of spatially structured populations. A first step towards greater practical applicability of the model with regard to conservation management is to express the rates c

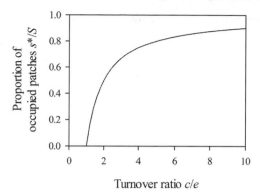

Figure 6.1 Steady state number of occupied patches s^*, scaled with respect to the total number of patches S, as a function of the turnover ratio c/e. The critical turnover ratio below which $s^*/S = 0$ is $(c/e)_c = 1$.

and e by parameters of the landscape. Hanski (1999) proposed to model the local extinction rate by

$$e = \frac{\varepsilon}{A^x},$$ (6.3)

where A is the size (carrying capacity) of the habitat patch harbouring the local population and ε and x are species–specific constants. Recalling that the extinction rate of a population is the inverse of the population's mean life time (Section 5.4), parameter x can be identified with the exponent $2r/\sigma^2 - 1$ of Eq. (5.18) where σ^2/r is the strength of environmental stochasticity.

The number of individuals emigrating from a local population, m, is likely to increase with the size of the local population, which is generally proportional to the size or carrying capacity A of the habitat patch. Depending on the species, the relationship between m and A may be concave, linear or convex:

$$m = \beta A^b,$$ (6.4)

where β and b are species–specific constants. Denoting the probability of a dispersing individual reaching another habitat patch as η (which may be determined by the distances between the habitat patches in the landscape and the dispersal ability of the species) and the rate by which an immigrant establishes a local population in an empty habitat patch as γ, the colonisation rate c is

$$c = \beta \eta \gamma A^b.$$ (6.5)

Inserting Eqs (6.3) and (6.5) into Eq. (6.2) delivers the equilibrium number of occupied habitat patches as

$$s^* = S \max\left[0, 1 - \frac{\varepsilon}{\gamma\eta\beta}A^{-(b+x)}\right]. \tag{6.6}$$

From this equation the critical patch size above which s^* is positive can be deduced to

$$A_c = \left(\frac{\varepsilon}{\eta\gamma\beta}\right)^{1/(b+x)}. \tag{6.7}$$

While the influence of the exponents of b and x are ambiguous, a clear conclusion from Eq. (6.7) is that the critical patch size increases with decreasing probability of successful dispersal among the habitat patches, η, and decreasing establishment probability γ.

Although Eqs (6.6) and (6.7) allow for assessing the viability of the metapopulation as a function of landscape (A, η) and species (e, x, b, β, γ) parameters, they still ignore the spatial characteristics of most real landscapes in which habitat patches may have different sizes A_i ($i = 1, \ldots, S$) and different pairwise distances between each other, so that $\eta = \eta_{ij}$ differs among pairs (i,j) of patches. Typically, the probability of a disperser reaching some patch j from some patch i is assumed to decline exponentially with increasing distance d_{ij} between the two patches:

$$\eta_{ij} = \exp(-d_{ij}/\delta), \tag{6.8}$$

where δ is the mean dispersal distance of the species.

To account for such spatial heterogeneity, Hanski and Ovaskainen (2000) replaced Eq. (6.1) for the number of occupied habitat patches with a set of equations for the probability p_i of observing a particular patch i ($i = 1, \ldots, S$) in the occupied state:

$$\frac{dp_i}{dt} = (1 - p_i) \sum_{j\,(j \neq i)} c_{ji}p_j - e_i p_i \tag{6.9}$$

with

$$c_{ij} = \beta\gamma\eta_{ij}A_i^b \tag{6.10}$$

and

$$e_i = \frac{\varepsilon}{A_i^x}. \tag{6.11}$$

The authors define the metapopulation as viable if, in the steady state, each habitat patch i has a non-zero probability of being occupied: $p_i^* > 0$ for all $i = 1, \ldots, S$. As a measure of metapopulation viability, the authors introduce the 'metapopulation capacity' λ_M, which is the leading eigenvalue of the matrix $\mathbf{M}$ with elements $m_{ij} = c_{ji}/e_i$ for $j \neq i$ and $m_{ij} = 0$ otherwise, and show that all the probabilities p_i^* are non-zero if

$$\lambda_\mathrm{M} > \frac{\varepsilon}{\beta\gamma n_{ij}}. \tag{6.12}$$

This condition is the 'spatial' equivalent of the viability condition $c > e$ in Fig. 6.1 and the positivity of the right-hand side of Eq. (6.6).

A remarkable observation in Eq. (6.12) is, however, that the viability of the metapopulation obviously does not depend on the number S of habitat patches in the landscape. This is highly implausible and explained by the fact that the model in Eq. (6.9) is deterministic in the sense that the rates c_{ij} and e_i are deterministic rates like the birth and death rates b and d in Eqs (4.2) and (4.3), and so the probabilities p_i change with time t in a deterministic manner. For the same reasons that birth and death rates are usually stochastic and must be interpreted as the probabilities of some (birth or death) events occurring within some small time step (Section 5.3), the rates c_{ij} and e_i should be regarded as stochastic rates. This implies that Eq. (6.9) should be replaced by a Markov chain model (similar to that in Section 5.4) from which the mean life time of the metapopulation can be calculated (Frank 2005). The formula derived by Frank (2005) in this manner for the mean metapopulation life time explicitly includes the number of habitat patches as a parameter, such that the mean metapopulation life time increases with increasing number of habitat patches. Other consequences of stochasticity in metapopulation dynamics are discussed in Frank (2005) and were outlined in Section 3.3.

Metapopulation models have many applications in spatial ecology (see, e.g., Section 8.4) and the analysis of spatially targeted market-based conservation instruments (see, e.g., Chapter 13). A question that has occupied the ecological literature for several decades and is still of interest is the question whether a population can better survive in a single large habitat patch or in a network of several small habitat patches (the 'SLOSS [single large or several small] debate'), or similarly, whether a few large patches are better than many small ones ('FLOMS' [few large or many small]') – under the constraint that the total habitat area is constant.

A rather comprehensive analysis of the issue has been carried out by Robert (2009) with the help of a metapopulation model that, in addition to the features described previously, includes spatially correlated environmental stochasticity. As described in Section 5.1, environmental stochasticity means that the survival conditions in the habitat patches vary randomly. If this variation is spatially correlated, then nearby patches exhibit more similar changes in the survival conditions than more distant patches. In the absence of spatial correlations, having several habitat patches may be beneficial because this allows for spreading the extinction risk so that if one habitat patch currently has poor conditions, another one may have good conditions. This advantage is reduced if the environmental variation becomes increasingly correlated so that all habitat patches have increasingly similar conditions.

The main results of the analysis by Robert et al. (2009) are that (i) the optimal number of habitat patches (recall: for given total area) increases with the level of dispersal among the patches; (ii) the optimal number of habitat patches first increases, and after some point decreases with increasing environmental variation, because on the one hand several habitat patches allow for spreading the extinction risk as noted, while on the other hand at very large and spatially correlated environmental stochasticity all patches have more or less the same conditions and it is better to have a single patch to avoid dispersal losses between patches; (iii) to maximise population survival, habitat patches should be arranged with some optimal distance between them that is large enough to escape the spatial correlation but small enough to keep dispersal losses (cf. Eq. 6.8) at an acceptable level and (iv) if a habitat network is in an optimal spatial configuration, the addition of habitat patches is most beneficial if the patches are added next to the existing network so the species' spatial range is increased, while a loss of habitat patches is more detrimental if the patches are taken out in a spatially correlated manner, reducing the species' spatial range.

The results indicate that the loss of spatial connectivity is on the one hand detrimental since it reduces the ability of individuals to reach other patches, but on the other hand can be beneficial if it allows for reducing spatial environmental correlations among the patches. Habitat fragmentation can therefore be beneficial in ecosystems subject to wildfires, or to combat the spread of a disease. An additional argument for habitat fragmentation is that it can facilitate the coexistence of species and thus may boost the level of biodiversity even if it is detrimental for individual species (Chapter 8).

The model by Robert et al. (2009) differs from the two previous metapopulation models by considering local dynamics, so that it not only distinguishes between occupied and empty habitat patches, but also explicitly considers the change in the size of a local population over time. This change is modelled as described in Chapters 4 and 5, adding the emigration and immigration of individuals. Although the SLOSS question can also to some extent be discussed without considering local dynamics, the neglect of local dynamics requires rather restrictive assumptions on the levels of environmental stochasticity and dispersal (Drechsler and Wissel 1997), and results are richer and more detailed if local dynamics are taken into account.

Another extension of the metapopulation models is to abandon the assumption that interactions between local populations through dispersal are only determined by the distance between the habitat patches (cf. Eq. 6.8). Instead, a habitat patch may have a link of any strength with another habitat patch. Such heterogeneity of patch interactions might be more likely in marine ecosystems in which dispersal is strongly influenced by water currents but can also be observed in terrestrial ecosystems (Fortuna et al. 2006). Some authors (e.g. Holland and Hastings 2008) have analysed the effect of habitat network topology on the dynamics of metapopulations. A research field in which network topology appears to be quite relevant is the spread of diseases (e.g. Oleś et al. 2013), which is affected not only by geographical distance but also by other factors such as transport of infected materials by humans. Altogether, the metapopulation concept is quite flexible and can be extended and applied to many problems in spatial ecology.

6.2 Grid–Based Models

In a grid-based model the landscape is structured as a grid with (usually square, sometimes hexagonal) cells. The state of each grid cell is defined by specific attributes, depending on the system under consideration and the model's purpose. Such cell attributes may be habitat quality for a species to reproduce and survive (which in turn may be a function of abiotic parameters such as temperature or precipitation, or biotic parameters such as presence of predators or prey), resistivity to dispersing individuals (dependent, e.g., on relief, vegetation cover, etc.) and more. Depending on its size, a grid cell may be identifiable as a patch for a local population, as in the patch-based models, or it may provide space for only a single individual such as a tree, or several adjacent cells may form a habitat patch for a local population or the territory of a single individual. As this description indicates, grid–based models are very flexible in representing different types of landscape structures.

A particular feature of grid-based models is that the state of a grid cell may change over time and that this transition between different states may depend not only on the current state of the focal cell but also on the states of the cells within some neighbourhood around the focal cell. Frequently used neighbourhoods in square grids are the so-called Moore neighbourhood of the eight adjacent cells and the von Neumann neighbourhood containing the four cells to the north, west, south and east.

In ecological grid-based models the state of a cell (e.g. occupied by a species or not) may change in accordance with the movement of individuals in the landscape (which in turn is affected by the cells' states). In these cases one has to distinguish between the grid cells and the individuals located in the cells. This type of model may be termed a grid-based individual-based model and will be covered in more detail in Section 7.2. If individuals are not considered explicitly but the focus is only on the cells and the transition rules, that is, the rules that describe how the cell states change from one time step to the next, the model is called a cellular automaton (Deutsch and Dormann 2005).

Next to the transition rules, for a grid-based model it has to be decided whether the states of the cells are updated between consecutive time steps in a synchronous or an asynchronous manner. In an asynchronous update each cell is considered one by one to update its state. This means that the first considered cell is updated according to the states of the cells in the current time step. In contrast, the second considered cell will be updated based on a landscape in which the first considered cell has already changed its state. Consequently, the transition of the first considered cell affects the transition of the second considered cell. This implies that the model dynamics depend on the sequence in which the individual cells are updated. Unless there are counterarguments provided by the properties of the focal system, to avoid bias the sequence of the updated cells is chosen randomly. To account for the implied stochasticity, the model dynamics are then simulated many times and the simulation outcomes evaluated statistically (cf. Chapter 5).

In a synchronous update, the new state is determined for each cell but stored in a 'cache', that is, the cell does not assume its new state immediately but the information about the new state is stored. Thus each cell – whenever it is considered in the sequence – 'observes' the current state of the cellular automaton but does not yet recognise the transitions of the other cells. Only after the new states have been determined for all cells and stored in the cache are they assigned to all cells at once. Consequently, the states of the cells in the new time step are independent of the sequence by which they were determined.

A second important decision in the design of a grid-based model – as in all spatial models – is whether boundary conditions are open, closed or periodic. Open boundaries assume that the system dynamics in the model landscape are imbedded in some larger system and affected by processes outside the model landscape, for example through the emigration of individuals into habitat patches or cells outside the model landscape. Closed boundary conditions assume that the system is isolated and the model landscape is confined by an impermeable 'wall' (here one can further distinguish between absorbing boundaries such that individuals moving 'into' the wall are absorbed and removed from the system, and reflecting boundaries where individuals moving 'into' the wall are sent back to the patch or grid cell from which they originated). Boundaries can have a considerable effect on dynamics and bias model results. To avoid this, one can set periodic boundary conditions, so that an individual that leaves the model landscape at one boundary (e.g. the eastern one) enters the system on the opposite boundary (the western one). By this the individual cannot leave the system nor any boundary and can, for example, move forward forever. Geometrically, if in a square grid boundaries are periodic in one dimension (e.g. in east–west direction), in three dimensions the model landscape looks like a cylinder, and if the boundaries are periodic in the other dimension (north–south direction) as well, the model landscape looks like a torus.

One of the first and most popular cellular automata is John Conway's 'Game of Life' (Gardner 1970). As outlined in Section 3.1, each grid cell of this cellular automaton is in one of two states, occupied by an individual or empty, and the transitions between these two states depend on the states of the cells in the neighbourhood. Considering the eight-cell Moore neighbourhood, Conway specified that (i) an occupied cell with more than three occupied neighbours becomes empty due to overpopulation and starvation; (ii) an occupied cell with fewer than two occupied neighbours becomes empty due to isolation and a higher risk of predation and (iii) an empty cell that is surrounded by exactly three occupied cells has ideal conditions for the species and becomes occupied through reproduction (and dispersal).

Simulating these dynamics for several time steps leads to patterns of the type shown in Fig. 6.2. The pattern changes over time but the 'population' is stable in the sense that it neither expands towards occupation of the entire landscape nor goes extinct where all cells are empty.

Figure 6.2 can be used to illustrate these transition rules. Single occupied cells, for example, will become empty in the next time step due to isolation (rule (i)). A straight horizontal line of three adjacent occupied cells (e.g. on the right end of the grid; note that boundary

Figure 6.2 Typical pattern of occupied (black) and empty (white) grid cells in Conway's Game of Life.

conditions are periodic) will turn into a straight vertical line of three adjacent occupied cells and vice versa, since the centre cell survives (rules (i) and (ii)), the two outer cells die from isolation (rule (i)), and the two empty cells next to the occupied centre cell become occupied (rule (iii)). A square of four occupied cells (middle of Fig. 6.2) remains unchanged, since each cell has three occupied neighbours while none of the surrounding empty cells has three occupied neighbours, so these remain empty. Experimenting with the transition rules reveals that the stable dynamics represented by Fig. 6.2 are the result of a delicate balance between the 'colonisation' of empty cells and the 'local extinction' of occupied cells.

The dynamics of Conway's Game of Life are affected not only by the bounds (in terms of the required numbers of occupied cells) specified in the transition rules but also by the rules' spatial dimension. The Moore neighbourhood represents a very local or short-ranged interaction. The opposite would be a long-ranged or even global interaction where the transition of a grid cell is affected by the states of the cells in a wider neighbourhood or all cells in the landscape. To demonstrate the influence of this interaction range, in the following I extend Conway's model and first generalise the transition rules by formulating that occupied cells survive if and only if the proportion of occupied neighbours exceeds about 0.2 (between the 1/8 that leads to starvation in Conway's Game of

Life and 2/8 that allows for survival) and is less than about 0.4 (between 3/8 and 4/8), while empty cells become occupied if and only if the proportion of occupied neighbours exceeds about 0.3 (between 2/8 and 3/8) and is less than about 0.4 (between 3/8 and 4/8).

Formulating the transition rules in terms or proportions rather than numbers of cells allows us to consider arbitrary sizes of neighbourhoods. I define the neighbourhood around some focal cell by a circle with radius R around the centre of that cell so that all cells with centres within that circle belong to the neighbourhood. By varying the radius from 1.5 to values of the order of the landscapes' spatial dimension, the neighbourhood can be expanded from the Moore neighbourhood to a global neighbourhood.

For reasons that will soon become apparent, I replace the crisp bounds within which survival and colonisation are possible by a probability function that specifies how the probabilities P_s and P_c of an occupied cell remaining occupied and an empty cell becoming occupied, respectively, depend on the proportion q of occupied cells in the neighbourhood:

$$P_k = \exp\{(q - \mu_k)^2/(2\sigma_k^2\}. \tag{6.13}$$

In this function, μ_k is the proportion of occupied cells that maximises survival ($k = s$) and colonisation ($k = c$), respectively, and σ_k measures how fast these probabilities decline if q deviates from μ_s and μ_c, respectively.

The model with the probabilities of Fig. 6.3 and various levels of the neighbourhood radius R is simulated for 1,000 time steps, using the first 50 time steps as a burn-in period to reach the steady state. Temporal means and standard deviations of the proportion of occupied cells and the average number of occupied cells in the Moore neighbourhood around an occupied cell are recorded. The latter state variable can be used to measure the level of spatial clustering, so that a small value indicates spatial scattering of occupied cells while a large value indicates strong spatial clustering.

Figure 6.4 shows that the mean proportion of occupied cells is rather unaffected by the neighbourhood radius R, but the variation of the proportion of occupied cells increases dramatically with increasing R. In addition, the mean level of spatial clustering increases with increasing R while the standard deviation of spatial clustering only increases slightly. This means that large neighbourhood radii lead to more clustered patterns and higher variation in the proportion of occupied cells over time.

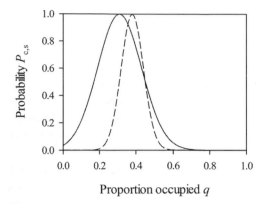

Figure 6.3 Probability P_s of an occupied grid cell staying occupied (solid line) and probability P_c of an empty cell becoming occupied (dashed line) as functions of the proportion q of occupied cells in the neighbourhood. Parameters for P_s: $\mu_s = 0.31$, $\sigma_s = 0.12$, and parameters for P_c: $\mu_c = 0.38$, $\sigma_c = 0.06$ (Eq. 6.13).

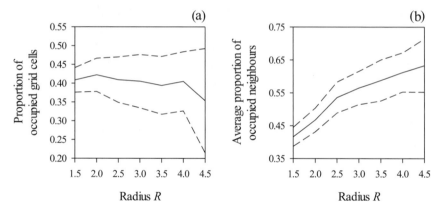

Figure 6.4 Proportion of occupied grid cells (left) and average proportion of occupied cells in the Moore neighbourhood around an occupied cell (right) as functions of the neighbourhood radius R. Solid lines: means; dashed lines: means plus/minus one standard deviation.

Similar to the stochastic populations in Chapter 5, such variation may eventually lead to the extinction of the population, and indeed, for neighbourhood radii of 5 or more the population does not survive the 1,000 time steps. The reason for the fluctuations at large R is that here all grid cells within some subregion (of magnitude R) experience the same conditions, so that more or less all *occupied* cells remain occupied or all become empty (depending on the jointly experienced proportion of occupied cells) and more or less all *empty* cells become occupied or all remain empty. In other words, the transitions between the cell states become more correlated in space as R increases, leading on the one hand to a higher level of spatial clustering but on the other hand to stronger fluctuations and eventually the extinction of the population. This effect is further aggravated if the transitions between the cell states are governed not by a smooth stochastic function but by crisp bounds as in Conway's original model, and global extinction occurs even at smaller neighbourhood radii R.

To conclude, the range of spatial interaction (from local 'direct-neighbour' type to global) can have an enormous impact on the dynamics of a spatially structured population, confirming the statement in Section 3.1 that the consideration of spatial structure (non-spatial with global interaction versus spatially explicit with local interaction) is an important feature of a model.

Although Conway's Game of Life represents a very artificial system, it contains many features of grid-based ecological models. Rather than listing further examples at this point, I refer to the grid-based individual-based models in Section 7.2 and the grid-based biodiversity models in Chapter 8. Furthermore, several of the land-use models in Chapter 13 are grid-based, as well as the ecological-economic models in Chapter 17, not to mention the various grid-based models already presented in Chapters 2 and 3.

Recent literature related to Chapter 6 includes Jackson and Fahrig (2016), Dhanjal-Adams et al. (2017), Gunton et al. (2017), Heinrichs et al. (2017), Rytteri et al. (2017) and Collard et al. (2018).

7 · *Models with Individual Variability*

In the previous chapters about ecological modelling, all individuals within a population were considered identical, so that they had, for example, the same probabilities of reproducing and surviving, and their spatial location in the habitat did not matter for reproduction or survival. This is, of course, a gross simplification of reality where individuals differ in their sexes and ages (implying differences in their reproduction and survival), their competitiveness in the struggle for resources, their behaviour and their spatial locations. In the present chapter I will present two approaches for the consideration of individual variability: equation-based models as presented in the previous chapters and individual-based models.

7.1 Equation-Based Models

One of the most obvious and oldest features (with regard to its consideration in models) of individual variability is age and its influence on reproduction and survival. The standard approach for the consideration of age is to consider discrete age classes and distinguish, for example, between newborns with ages between zero and one year, individuals with ages between one and two years, and so forth. All individuals within the same age class have the same survival probability and produce the same number of offspring within a given unit of time, but between age classes these parameters may differ. The analysis of such an age-structured population is simplified by assuming that the population dynamics run in discrete time steps of length Δt. In discrete time the ageing and survival of individuals can be described by

$$n_j(t + \Delta t) = p_j n_j(t) + s_{j-1} n_{j-1}(t) \quad j > 1, \tag{7.1}$$

where $n_j(t)$ is the number of individuals in age class j, p_j is the probability of an individual in age class j surviving to the next time step $t + \Delta t$, and s_j

is the probability of an individual in age class j surviving to the next period *and* entering the next age class. Thus, Eq. (7.1) describes that from n_j individuals in age class j an (average) proportion p_j survives and stays in the same age class, while from n_{j-1} individuals in age class j–1 a proportion s_{j-1} survives and ages, and both numbers together form the number of individuals in age class j in the next time step.

If f_j is the (average) number of offspring that is produced per individual in age class j and survives till the end of the current time step, then the number of newborns in the next time step is

$$n_1(t+\Delta t) = f_1 n_1(t) + f_2 n_2(t) + \ldots = \sum_j f_j n_j(t). \qquad (7.2)$$

Equations (7.1) and (7.2) can be written in matrix form

$$
\begin{bmatrix}
n_1(t+\Delta t) \\
n_2(t+\Delta t) \\
\vdots \\
n_{J-1}(t+\Delta t) \\
n_J(t+\Delta t)
\end{bmatrix}
=
\begin{bmatrix}
f_1 & f_2 & \cdots & f_{J-1} & f_J \\
s_1 & p_2 & \cdots & 0 & 0 \\
0 & s_2 & \ddots & \vdots & \vdots \\
\vdots & & \ddots & p_{J-1} & 0 \\
0 & \cdots & 0 & s_{J-1} & p_J
\end{bmatrix}
\times
\begin{bmatrix}
n_1(t) \\
n_2(t) \\
\vdots \\
n_{J-1}(t) \\
n_J(t)
\end{bmatrix},
\qquad (7.3)
$$

where the numbers of individuals in the various age classes in the next time step, $n_j(t + \Delta t)$, are arranged in a vector that is obtained by multiplying a vector with the current number of individual, $n_j(t)$, with a transition matrix (Tuljarpurkar and Caswell 1997). This matrix is called a Leslie matrix (sometimes also referred to as a Levkovitch matrix, to reserve the term Leslie matrix for the special case of a zero diagonal, $p_j = 0$) (Caswell 2001). The Leslie matrix in Eq. (7.3) contains in the first row the fecundities f_j, in the diagonal the probabilities of surviving and not ageing, p_j, and in the subdiagonal below the diagonal the probabilities of surviving *and* ageing, s_j. A maximum of J age classes is considered.

Equation (7.3) allows us to use linear algebra to deduce characteristic features of the population dynamics. If the Leslie matrix has certain properties (such that each age class affects at least one other age class), after a sufficient number of time steps the population dynamics reach a steady state in which the total population size $N = n_1 + n_2 + \ldots n_J$ multiplies in each time step by a constant factor λ. This factor can be shown to be the largest non-zero and real eigenvalue of the Leslie matrix. In addition, after the transition time the proportions of individuals in the particular age classes, n_1/N, n_2/N, $\ldots$, n_J/N, are constant in time and

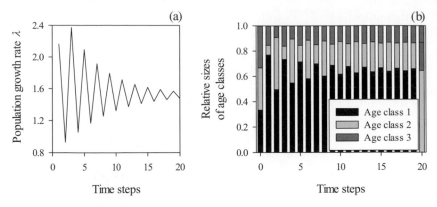

Figure 7.1 Population growth rate (left) and relative sizes of the age classes (right) as functions of time. The number of age classes is $J = 3$; the fecundities are $f_1 = 0$, $f_2 = 4$, $f_3 = 1$; the survival/transition rates are $s_1 = 0.5$, $s_2 = 0.9$, $p_3 = 0.1$; all other entries in the Leslie matrix are zero. The initial population structure is $n_1(0) = n_2(0) = n_3(0) = 10$ and the length of one time step is $\Delta t = 1$.

given by the elements of the right-hand eigenvector associated with eigenvalue λ (Caswell 2001).

An example is shown in Fig. 7.1 where in the steady state the population grows at a rate of $\lambda \approx 1.5$ and the relative sizes of the age classes are $n_1 : n_2 : n_3 = 0.65 : 0.21 : 0.14$. The example also demonstrates that the transient time required to reach the steady state from some initial state may be quite long. In small populations, for example, after the reintroduction of a species into a new habitat, this can be critical for the success of the conservation project.

Similar to a homogenous population, the fecundity and survival rates in an age-structured population may be stochastic. The matrix approach described previously allows for taking such stochasticity into account (Fieberg and Ellner 2001). Even density-dependence (carrying capacities) can be considered, but then many of the computational advantages of the matrix formulation are lost.

Next to age, another relevant source of variability is affiliation to a particular species. Although multi-species models are usually not discussed in the context of variability, from a formal point of view it makes little difference whether two individuals in a habitat differ because of their ages or other attributes or because they belong to different species (except, of course, that individuals from different species do not mate with each other).

Classical models of species interactions are found in Lotka (1920) and Volterra (1931) for the description of the joint dynamics of two species.

The model for a predator–prey system in which one species feeds on the other has already been presented in Section 3.5. For two competing species, numbered 1 and 2, the joint population dynamics can be described by

$$\frac{dN_1}{dt} = r_1\left(1 - \frac{N_1 + \beta N_2}{K_1}\right)$$
$$\frac{dN_2}{dt} = r_2\left(1 - \frac{N_2 + \gamma N_1}{K_2}\right). \tag{7.4}$$

For each species the population dynamics are described by logistic growth, determined by a species-specific carrying capacity K_i as in Eq. (4.10). In addition, the growth of each species is limited not only by the number of conspecifics (the first term in each fraction) but also by the number of individuals from the other species ($\beta N_2/K_1$ and $\gamma N_1/K_2$, respectively). The coefficients β and γ determine how strongly an individual from the other species affects the growth compared with an individual from the same species.

An interesting question is under what conditions the model of Eq. (7.4) leads to coexistence where both species have a positive population size. In a deterministic model such as the one in Eq. (7.4), it is appropriate to answer this question by searching for the steady state in which $dN_1/dt = dN_2/dt = 0$. Applying this procedure to Eq. (7.4) and solving for the equilibrium population sizes of the two species, we find that both are positive if

$$\gamma/K_2 < 1/K_1$$
$$\beta/K_1 < 1/K_2. \tag{7.5}$$

The first inequality states that the inhibitory influence of species 1 on species 2 (γ/K_2) is smaller than that of species 1 on itself ($1/K_1$), and the second inequality states the same for species 2 (Begon et al. 1990, ch. 7.4). So in other words, the two species coexist if the interspecific competition between the two species is weaker than the intraspecific competiton within each species. For a case in which Eq. (7.5) is fulfilled, Figure 7.2a shows exemplary population trajectories where the steady state with both species present is reached independent from the initial population sizes.

If the first inequality of Eq. (7.5) is reversed and the inhibitory influence of species 1 on species 2 becomes too strong, then species 2 becomes extinct (Fig. 7.2b, solid line), while reversal of the second inequality implies the extinction of species 1 (dashed line). If both inequalities are reversed so that the inhibitory influences of both species on the

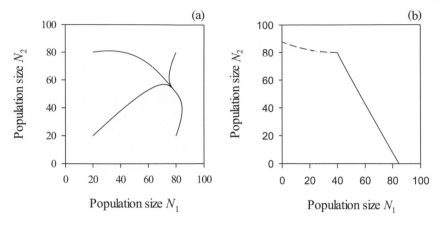

Figure 7.2 Population trajectories for the model of Eq. (7.4) for various parameter combinations and initial conditions. Panel a: $\beta = 0.4$, $\gamma = 0.6$, $K_1 = K_2 = 100$, $r_1\Delta t = r_2\Delta t = 0.5$; several initial conditions. Panel b: $\beta = 0.5$, $\gamma = 1.5$ (solid line) and $\beta = 1.5$, $\gamma = 0.5$ (dashed line) with initial condition $N_1(0) = 40$, $N_2(0) = 80$. Carrying capacities and population growth rates are as in panel a.

respective other species exceeds that on their own conspecifics, a steady state with positive population sizes exists but is unstable.

Generally one can say that if two or more species compete for the same resource, the strongest competitor outcompetes the weaker competitor(s). Coexistence is possible only if the species occupy different niches, such as utilising different temperature ranges or foraging at different times or places. This rule, known as the Competitive Exclusion Principle (Begon et al. 1990, ch. 7.4), implies that biological diversity relies heavily on spatial and temporal diversity of the species' living conditions. Chapter 8 will provide further examples where models are used to better understand the causes of biological diversity.

7.2 Individual-Based Models

In principle, equation-based models can also be used to consider other dimensions of individual variability. However, if several dimensions of variability are considered, each with several levels (such as 'low', 'medium', 'high'), the total number of classes, that is, combinations of levels on the various dimensions, would be so high that on the one hand the model would become intractable and on the other hand the number of individuals in each class might become so small that transitions between classes could not be modelled in a meaningful manner.

In these cases it is more appropriate to 'reverse' the structure of the model and, instead of considering classes and recording how many individuals are found in each class, consider individuals and record their properties (age, fitness, location, etc.). These models are called individual-based models and have become a prominent tool in ecological research in recent decades, as well as in economic research (where they are referred to as agent-based models: cf. Chapter 12) and other scientific disciplines (Grimm 1999; Grimm and Railsback 2005; Railsback and Grimm 2012).

To demonstrate that there is an equivalence of equation- and individual-based modelling approaches, reconsider the previous example of an age-structured population (Eq. (7.3) and Fig. 7.1). The individual-based version of that model is to consider a number of individuals, each known by name (or some index) and age. In each time step, for each individual of a given age the number of offspring is calculated, and it is decided whether the individual survives or not. Since survival rates lie between zero and one, survival in the simulation is modelled stochastically: if some random number drawn from the interval $[0,1]$ is below $s_i + p_i$ (cf. Eq. 7.1) the individual (of age i) survives, and if in addition it is below s_i the individual ages and is assigned an age of $i + 1$. Similarly, the number of offspring is drawn from the Poisson distribution with mean f_i. Dying individuals are erased from the list of names (indices) and newborn individuals are added with new names (indices) and given an age of one. In the following time step this new set of individuals is treated in the same manner as the individuals in the previous time step, and this procedure is continued for all subsequent time steps of interest. The resulting dynamics obtained for the parameter values of Fig. 7.1 are summarised in Fig. 7.3. One can observe an extraordinary agreement with Fig. 7.1.

As previously noted, while in a population that is only structured by the individuals' ages an equation-based model is likely to be more efficient than an individual-based model, if more dimensions of variability are added the individual-based approach is most likely to be more appropriate. Features that can hardly be considered in equation-based models include (Grimm et al. 2006):

1. multiple sources of variability in individual characteristics,
2. implied by (1), complex life cycles of the individuals and social structure of the population,
3. locations, spatially heterogeneous environments and local interactions among individuals and
4. adaptive behaviour, including autonomous pursuit of certain objectives, sensing, learning, foresight, adaptation and more.

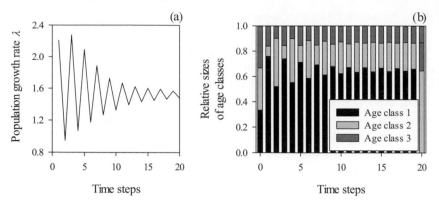

Figure 7.3 Population growth rate (left) and relative sizes of the age classes (right) as functions of time as obtained from an agent-based simulation model (see text). Demographic parameters and initial condition are the same as in Fig. 7.1. The results are averages over 100 simulation replicates to account for the stochasticity in the simulation.

An example that fulfils all these four criteria is the model by van Winkle et al. (1998) for trout species. The model combines a physical model of river streams with a population dynamics model to allow for population-level assessments of the effects of stream flow and temperature regimes on the trout. The model is dynamic on daily time steps and spatially differentiated and explicit, since habitat quality, for example, depends on the spatial location (e.g. water depth) in the stream, and the inter-action among the fishes depends on the fishes' relative spatial locations. The model simulates the complete life cycle of the trout species, from egg to adult.

On the physical side, the model simulates the velocity of the water and the water depth as a function of the geometric structure of the river and the water flow. Individual trout select their spatial location in the stream based on certain habitat characteristics such as water velocity, depth and density of other trout. In the selection of their locations, trout compete for two resources: feeding station and cover. In this competition, larger trout have an advantage over smaller trout, which may suffer energetic losses if they cannot find a preferred feeding station.

Each day a trout may move, and it does so based on two objectives: minimisation of mortality and maximisation of growth (a more recent model of brown trout shows that this objective function does not yet allow for capturing all realistic behaviours: Railsback and Harvey (2002)). A trout's expectation of mortality risk and daily growth in weight is simulated as a running average of the ratio of its unique history of mortality risks and growth, weighted by a memory factor. Reproductive success (number of eggs, developmental rates and probabilities, etc.) depends on the size of the female and on water temperature. The growth of an individual trout depends on the trout's balance of energy acquisition and energy costs, which depends on the trout's behaviour and the local physical conditions, including the amount of food.

For energy acquisition, trout choose between two strategies: 'sit-and-wait' and 'active'. Prey consumption is modelled depending on prey encounter, which is decisively influenced by water velocity; probability of attack; and probability of capture. Energetic costs include daily respiration and activity respiration. These energetic costs again depend on the trout's behaviour, in particular the choice between locations with higher cover versus higher food abundance, and the type of foraging behaviour. Mortality of trout is determined by various factors including water temperature and flow, predation and others, and thus emerges from the decisions and behaviour of the individuals rather than being imposed via observed or postulated mortality rates. This emergence from adaptive behaviour is a key characteristic of fully fledged individual-based models, as it allows us to let the response of a population to new conditions, for example warmer water, floods or high turbidity, emerge from behavioural responses. Basing individual-based models on so-called first principles from which population responses emerge is a relatively recent trend in ecological modelling. This approach is more complex than traditional modelling approaches, but it is more flexible and in many instances more useful (Stillman et al. 2015). Examples of such first principles in ecology include fitness seeking and energy budget as in the trout example, as well as physiology, photosynthesis and stoichiometry.

Thus the consideration of behaviour is probably the most important argument for an individual-based model. As in the example of the trout model, it includes the question of how individuals sense their environment (e.g. water velocity), what their decision-making objectives are (e.g. maximisation of fitness determined by reproductive success and longevity), prediction (e.g. how will fitness depend on the choice of the spatial location), adaptation (e.g. how to adapt to changing water velocity), interaction (e.g. competition with other trout for feeding stations) and social structure (here simply considered as a competition hierarchy from the largest to the smallest individual). These behavioural characteristics are listed as 'design concepts' in the ODD (overview, design concepts and details) protocol for the description of individual-based models by Grimm et al. (2006) and also represent important features in (human) agent-based models (Chapter 12).

Behaviour is an important component of many movement models that describe how individuals move in spatially structured landscapes. Often it is not easy to detect such movement rules from observed individual movements in a landscape. An example of an individual-based model with the objective of detecting such movement rules is Pe'er et al. (2013), which considers so-called hilltopping movement in butterflies. Males and females of some butterfly species aggregate on the tops of hills to increase their chance of finding a mate. The authors compare various butterfly movement rules with regard to the maximisation of mating success. The movement rules are defined by two probabilities: the probability of maintaining the current direction of flight; and in case the flight direction is changed, the probability of choosing the highest positive slope locally available. Analysis of the model reveals that mating success is maximised if the butterflies have a non-zero probability of maintaining the current flight direction and/or if the probability of flying uphill is strictly below one. The plausible argument for this result is that this movement behaviour of not always choosing the steepest positive slope reduces the risk of becoming trapped in a local maximum, that is, a hilltop that overtops the immediate surroundings but has a lower elevation than other hilltops. This behaviour is very similar to the functioning of the simulated annealing algorithm for the numerical solution of complex optimisation problems (Section 2.7).

If one assumes that the movement rules of the butterflies have evolved to maximise mating success, one can conclude from the analysis of Pe'er et al. (2013) that butterflies of the considered species disperse in the described manner with a non-zero probability of maintaining the current flight direction and below-one probability of flying uphill.

A different approach for detecting movement behaviour of animals is pattern-oriented modelling (cf. Section 2.6), where model output is compared with observed patterns, and model assumptions that lead to the highest agreement between model output and observation are most likely to be valid for the modelled system. Eichhorn et al. (2012) simulate the foraging behaviour of a red kite (*Milvus milvus*) around its eyrie to estimate the bird's risk of collision with a wind turbine as a function of the distance between eyrie and wind turbine. The movement behaviour of the red kite is described by model parameters whose values are largely unknown. Comparing the model output with field observations enables the authors to draw conclusions on which movement behaviour is most likely to apply for the species and establish a 'damage function' that predicts the bird's annual mortality as a function of the distance between eyrie and wind turbine.

While the movements of the butterflies in Pe'er et al. (2013) and the red kite in Eichhorn et al. (2012) are induced by the landscape structure and the individuals' behavioural responses to that structure, collective movements can occur in an entirely self-organised manner through interactions between individuals. A popular example of this is the formation of ant trails, where ants deposit pheromones which are then followed by other ants. Watmough and Edelstein-Keshet (1995) developed an agent-based model for trail formation in ants. In their model, moving ants deposit pheromones that decay with time. An ant sensing the pheromone has a certain probability of following the trail which increases with increasing pheromone concentration. If an ant has a choice between two trails, it chooses the trail with the higher pheromone concentration. In the course of the simulation, trails form. Among other things, the authors find that the strengths of the trails, that is, the pheromone concentration per unit length, increase with increasing fidelity of the ants, that is, with increasing probability of an ant remaining on the trail. There is a positive feedback, so that a higher pheromone concentration leads to higher fidelity and more ants following the trail, which in turn increases pheromone concentration. Counter-intuitively, a higher level of fidelity leads to the formation of fewer strong but a higher number of weak trails. The reason for this is that higher fidelity implies that ants stay on a trail even if it is weak, leading to the formation of many weak rather than few strong trails. These dynamics of trail formation are an instructive example of how behaviour and interaction among individuals (even if indirect via pheromones as in the present case) can lead to the emergence of complex spatio-temporal patterns.

Recent literature related to Chapter 7 includes Zeigler and Walters (2014), Thiery et al. (2015), Lonsdorf et al. (2016), Radchuck et al. (2016), Regan et al. (2017), Chen et al. (2018), Grant et al. (2018), Přibylová (2018), Rinnan (2018), Rueda-Cediel et al. (2018) and Warwick-Evans et al. (2018).

8 · *Models of Biodiversity*

The previous four chapters presented different types of ecological models considering important features of ecological systems such as intraspecific competition and density-dependence, demographic and environmental stochasticity, patch- and grid-based spatial structure and individual variability and behaviour. In the present chapter I will demonstrate how these modelling approaches can be joined and used to explore causes and consequences of biological diversity, that is, the question of why different species can coexist in the same region and what this implies, for example, about the resilience of ecosystems in the face of disturbances and change.

8.1 Niche Separation and Resource Partitioning

The first cause of biodiversity has already been discussed briefly in Section 7.1: if intraspecific competition is stronger than interspecific competition. In the model of Eq. (7.4) two species rely on two resources, and if both species have a smaller competitive effect on the respective other species than on their conspecifics they can coexist. This condition is very much related to the niche, or resource partitioning, concept. Simply speaking, a niche comprises the range of values of biotic and abiotic parameters such as climatic conditions, light, presence of other species and others within which the species can survive and reproduce. If two species occupy different niches they can coexist even if one of the species is more competitive than the other in terms of longevity and/or reproductive success. Empirical examples of this biodiversity concept can be found, for instance, in Begon et al. (1990, ch. 7). Other explanations of biodiversity will be presented in this chapter by means of modelling examples. A generalised version of the niche concept is the so-called modern coexistence theory by Chesson (2000) which identifies, under certain simplifying mathematical assumptions (Barabás et al. 2018), two mechanisms promoting coexistence: 'stabilising', where species at low abundances have a competitive advantage because they are released from intraspecific competition, and 'equalising', where average fitness differences between species are reduced, for example due to resource limitation. The latter mechanism can only slow down competitive exclusion, whereas the former can lead to long-term coexistence.

8.2 Spatial Structure and Stochastic Disturbance

A second cause of biodiversity that was proposed some decades ago is disturbance and the resulting spatial heterogeneity in landscapes. Levin and Paine (1974) regard an ecological community as 'a spatial and temporal mosaic of . . . small scale systems, recognising that the individual component islands or 'patches' cannot be viewed as closed. . . . the overall patterning must be understood in terms of a balance reached between extinctions and the immigration and recolonisation abilities of the various species.'

In their model, disturbance such as fire, drought, flood and so forth randomly destroys patches of habitat and creates new habitat patches elsewhere (or alternatively, it destroys the local populations inhabiting the habitat patches and thereby creates empty patches). Thus, disturbance has two effects: it creates open spaces for colonisation and it prevents the communities in the individual patches from reaching an equilibrium in which only the strongest competitor prevails.

To demonstrate how spatial structure and disturbance can ensure the coexistence of multiple species, Levin and Paine (1974) presume that the species composition in a habitat patch is largely determined by the age a of the patch, that is, the time since it was created or the time since all existing local populations in the patch were destroyed. If disturbance can be shown to create a mosaic of patches with different ages, one can conclude that it facilitates biodiversity.

To model the impact of disturbance on patch age distribution Levin and Paine (1974), consider an ensemble of patches with different ages a and sizes ξ. The model is spatially implicit in that the spatial arrangement of the patches is ignored. The authors define a density function $n(t,a,\xi)$ for patches with age a and size ξ at time t, so that $n(t,a,\xi)\Delta a\Delta\xi$ is the number of patches with ages between a and $a + \Delta a$ and sizes between ξ and $\xi + \Delta\xi$. If $g(t,a,\xi)$ is the mean growth rate of patches of age a and size ξ at time t, and $\mu(t,a,\xi)$ is the extinction rate of patches of age a and size ξ at time t, the dynamics of the system can be described by a partial differential equation for n:

$$\frac{\partial n}{\partial t} + \frac{\partial n}{\partial a} + \frac{\partial (gn)}{\partial \xi} = -\mu n. \tag{8.1}$$

Equation (8.1) can be solved for particular assumptions regarding the functional shape of the growth rate $g(t,a,\xi)$. Depending on the rate by which new patches of age zero are created (or equivalently, patches occupied with local populations become empty), different levels of heterogeneity in patch ages and sizes, and hence different levels of biodiversity, can be generated.

8.3 Endogenous Disturbances

In Section 8.2 disturbance was an exogenous process that destroyed habitat patches and created new ones. In addition, disturbance can also be an endogenous process. Wildfires, for instance, are initiated by exogenous factors such as lightning, but whether a forest starts to burn and if and how the fire spreads to neighbouring stands depends on the current structure of the forest, and this is endogenous. Some models of wildfire spread were touched on briefly in Section 2.2. In the present section I will focus on one of these models, the Drossel–Schwabel model (Zinck et al. 2010).

In the Drossel–Schwabel model the forest is described by a grid of square cells, each of which can be occupied by a tree or not. Empty cells regrow with probability p, and occupied cells burn (and become empty) at a probability $f \ll p$. The four neighbours of a burning cell will also burn. Zinck et al. (2010) modified the Drossel–Schwabel model by assuming that a grid cell represents a forest stand rather than a single tree, and introduced the probability of a cell having burnt a number of a time steps ago. This probability can be calculated as

$$P(a, q) = 1 - (1 - q(1 - T))^{a}, \tag{8.2}$$

where q is the average proportion of grid cells with trees. Lastly, the authors related the 'age' a of a stand to its successional state and measured the diversity of successional states in the forest by the Shannon–Wiener index (Begon et al. 1990, ch. 17.2.1). One of the results produced by simulating the model is a relationship between disturbance level (annual area of forest burnt) and diversity of successional states. This relationship turned out to be hump-shaped, so that the diversity was maximised at intermediate levels of disturbance. Such a non-monotonic dependence of diversity on disturbance has also been observed in a real forest with ecological parameters similar to those of the modelled forests (Zinck et al. 2010) and is referred to as the Intermediate Disturbance Hypothesis (cf. Section 8.7).

Another popular example of endogenous disturbance is gap dynamics in forests. Starting from a mature forest, old trees die now and then and fall down. By this they create gaps of open space with bare ground and much light. This creates opportunities for pioneer species that need and tolerate light to germinate and grow. In their shade, other – more competitive – tree species become established and with time overgrow the pioneer species, which eventually decline. In the end the competitive trees prevail – until another of these trees dies and falls down. Therefore the forest is a mosaic of older and younger stands with different associated tree species.

Essentially, this mosaic is comparable to the pattern of habitat patches produced by the model in Section 8.2, but it is created not by an exogenous disturbance but by the species – the trees – themselves. Of course, exogenous disturbances such as storms and fires can influence this mosaic cycle, but the cycle also runs without these exogenous disturbances.

A model that investigates the interplay of endogenous and exogenous disturbances in a beech forest is Rademacher et al. (2004). The model, named BEFORE, is grid-based with grid cells of size 14 by 14 m^2, which is about the canopy size of a large beech tree. Four vertical layers or tree size classes are considered: seedlings, juveniles, and small and large adults. Survival and the transition between age classes are described by equations (cf. Section 7.1) for seedlings and juveniles, and in an individual-based manner (cf. Section 7.2) for adults. A simulation time step covers 15 years.

The forest dynamics are described by a number of rules. In a closed canopy (highest layer of large adults), larger trees grow at the expense of smaller trees. If the canopy is not closed, smaller trees from the upper layer grow, or small adults from the layer below fill the space. Small adults can grow into large adults only if they have remained as small adults for some time; after a while, if they do not get the chance to grow into large adults they die. Growth of juveniles into small adults depends on the amount of light received, which depends on the canopy cover. Seedlings are shade-tolerant and grow into juveniles at a particular rate, independent of the canopy cover. Regeneration, that is, the establish-ment of seedlings, depends on the presence of adults. Adults age and die at a rate dependent on their age. Available light depends on the conditions not only in the focal cell but also in the neighbouring cells. If a tree falls due to a storm it can damage the canopy in downwind cells.

The model renders the mosaic cycle described previously: a reduction of the cover in the upper canopy leads to an increase of the cover in the lower layers. The trees of these layers grow and eventually close the upper canopy. This in turn reduces the light conditions for the lower layers, reducing their cover. When the upper canopy opens due to the death of a large adult, the cycle starts again.

The described temporal dynamics are associated with spatial dynamics. These depend on the presence of storms as an external disturbance. Without the consideration of storms, the dynamics of all grid cells in the forest are synchronised, but if storms are included, the dynamics are synchronised within clusters of adjacent grid cells and desynchronised between different clusters. The reason is that the storms lead to the random tipping of individual trees, which affects the dynamics in the neighbouring grid cells, but this interaction proceeds only to a limited distance, which forms the boundary of the cluster.

Within the context of biodiversity one has to state that the model BEFORE considers only a single species – beech – but one can argue that the different life stages of the beech trees are associated with other forest species. In addition, the same modelling approach can be applied to forests with multiple tree species as well.

An example of such a multi-species forest model is FORMIND by Köhler and Huth (1998), which has been used to model rainforests in many continents and regions. In the model, the large number of tree species present in rainforests is classified into 22 functional groups, ranging from shade-tolerant to shade-intolerant species. The shade-intolerant species are pioneer species that establish themselves in forest gaps created by falling trees, while the shade-tolerant species establish themselves only below a closed canopy but eventually grow into the upper canopy of the forest. Each individual tree is known by its size. Similar to BEFORE, the forest is structured into a grid of square cells, each of the canopy size of a large tree.

Trees only have a single state variable, stem diameter at breast height, and the corresponding biomass. All other structural tree features are derived via allometric relationships. The growth of each individual tree is described by a differential equation, taking into account the processes of photosynthesis, losses for deadwood, respiration and below-ground growth. The growth model further considers tree geometry, light condition, competition for light, species- and size-dependent mortality and recruitment. Due to its detailed consideration of photosynthesis and hence carbon fixation, the model is well suited, for example, to describe the impact of rainforests and their structure on the global carbon cycle (Dantas de Paula et al. 2015; Brinck et al. 2017).

The interaction of endogenous and exogenous disturbances also affects the coexistence of trees and grass in savannahs (Jeltsch et al. 1996, 2000). The model by Jeltsch et al. (1996) is again grid-based with each grid cell having a size of 5 by 5 m^2, which corresponds to the size of a large tree. The model distinguishes, among others, between tree-dominated cells, shrub-dominated cells, cells dominated by perennial grasses and herbs, cells dominated by annuals and empty cells. Soil humidity can take four different levels in two soil layers and affects the establishment of seedlings and the growth of vegetation. Fire and browsing by animals prevent the savannah from changing into forest (Jeltsch et al. 2000): an increased proportion of trees reduces the risk of fire but increases the number of browsing animals while reducing the number of grazing animals. This feedback altogether facilitates fires and conserves the grass species in the savannah. The transition to grassland, in turn, is prevented by microsites in which trees can become established. These microsites are caused by heterogeneities in soil or relief and by termite mounds (Jeltsch et al. 2000). The interaction of these factors and disturbances altogether prevent the savannah from turning into grassland or forest and leads to the coexistence of tree, shrub and grass species.

8.4 Trade-Offs between Species Traits

Next to disturbances and resource partitioning, a third factor facilitating coexistence are trade-offs between species traits, in particular between competitive strength and the ability to disperse and colonise empty habitat patches (Tilman 1994). The author argues that if the inferior competitor is the better disperser or coloniser (known as the 'competition–colonisation trade-off') it is able to occupy parts of the landscape that are (currently) not occupied by the superior competitor. The justification for this statement is that due to local extinctions a species can never occupy all habitat patches in a landscape but only a proportion

$$p^* = 1 - \frac{e}{c} \tag{8.3}$$

of it, where c and e are the colonisation and extinction rates, respectively (cf. Eq. (6.1) in Section 6.1). Every time a habitat patch becomes empty due to a local extinction of the superior species, the inferior species with its higher colonisation ability has a chance of occupying the new empty patch, and will have the chance to persist locally until the superior species colonises this patch again.

To support these verbal arguments, Tilman (1994) extended the single-species metapopulation model of Levins (1969) (Section 6.1) to a two–species model where the dynamics of the superior species 1 are described by the same equation as Eq. (6.1):

$$\frac{dp_1}{dt} = c_1 p_1 (1 - p_1) - e_1 p_1 \tag{8.4}$$

with p_1 being the proportion of habitat patches occupied by the superior species and c_1 and e_1 the colonisation and local extinction rates of that species. In contrast, the inferior species 2 can colonise only habitat patches occupied neither by species 1 nor by species 2, reducing the colonisation term of the metapopulation equation from $c_2(1 - p_2)$ to $c_2(1 - p_1 - p_2)$; and in addition, if species 1 colonises a patch occupied by species 2, this provokes the local extinction of species 2, leading to the additional extinction term $c_1 p_1 p_2$:

$$\frac{dp_2}{dt} = c_2 p_2 (1 - p_1 - p_2) - e_2 p_2 - c_1 p_1 p_2. \tag{8.5}$$

Both species can be shown to coexist with positive equilibrium occupancy p_1^* and p_2^* if

$$c_1 > e_1 \quad \text{and} \quad c_2 > \frac{c_1 (c_1 + e_2 - e_1)}{e_1}. \tag{8.6}$$

If both species have the same local extinction rate, $e_1 = e_2$, then due to the first inequality, $c_1 > e_1$, the second inequality is fulfilled for $c_2 > c_1$, that is, if the inferior species 2 has a higher colonisation rate than the superior species 1. If species 2 has a smaller local extinction rate than species 1, $e_2 < e_1$, even smaller colonisation rates c_2 suffice for species 2 to persist in the landscape together with species 1.

8.5 Neutral Theory

A completely different theory for the existence of biodiversity is to ignore differences among the individuals of different species altogether but explain their coexistence by random processes. In the Unified Neutral Theory of Biodiversity and Biogeography by Hubbell (2001), all species are assumed to have identical rates of reproduction and mortality and identical dispersal abilities. An ecosystem is regarded as a metacommunity with J_M individuals that consists of a number of local communities, each containing J individuals. If an individual in a particular local community dies, it is replaced with probability $1 - m$ by an individual with random species identity from the same local community, and with probability m from another local community. This assumes that dispersal of individuals among different communities is possible but limited.

In addition, a replaced individual is assigned a new species identity (so-called point mutation) with probability v. Introducing the fundamental biodiversity number $\Theta = J_M v/(1 - v)$, the species richness, that is, the number of species, in a local community can be calculated as

$$S = \sum_{k=1}^{J} \frac{\Theta}{\Theta + k - 1} \tag{8.7}$$

(Rosindell et al. 2011). In addition, the distribution of the abundances of the different species can be calculated. Comparing the model predictions of species richness and species abundances with empirical data has led to mixed results (Rosindell et al. 2011; cf. Section 8.6) but has often worked surprisingly well. This has led to considerable controversy as it challenges a view which has been taken for granted by most ecologists, that niche differences are essential to explain diversity in ecosystems. Altogether, the neutral theory is a good null model against which alternative explanations of biodiversity can be compared but it has limited capability to explain biodiversity patterns in real-world ecosystems.

8.6 Neutral Theory with Trade-Offs

While the neutral model of biodiversity explains species-abundance distributions and species-area relationships relatively well, it appears too simplistic for capturing more detailed patterns of biodiversity. Wiegand et al. (2017) compared the neutral model with alternative models in which demographic rates for reproduction and mortality vary. In their analysis that variation was based on field data from a rain forest in Panama. The authors simulated the dynamics of a 50 hectare rainforest plot, consisting of 20 by 20 m^2 quadrats, and tested the predictive abilities of the different models with regard to a number of criteria, including the number of trees and the number of species per quadrat, and the correlation between number of trees and number of species.

The criteria were measured for each model variant and compared with the field data from the rainforest plot. Compared with the neutral model, the alternative model variants with variable demographic rates reproduced the observed patterns with higher accuracy. As the authors explain, spatio-temporally varying biotic conditions, for example increased recruitment rates in tree-fall gaps, have an important effect on the biodiversity in rainforests.

Another test of the neutral theory was performed by dos Santos et al. (2011), who analysed whether a neutral model is able to reproduce the Intermediate Disturbance Hypothesis (Connell 1978), which states that in a disturbed ecosystem diversity is maximised at intermediate levels of disturbance (cf. Section 8.2). The plausible explanation for this pattern is that at very low disturbance levels the competitively superior and disturbance-intolerant species dominate, while at high disturbance levels the competitively inferior and disturbance-tolerant species dominate; meanwhile, at intermediate disturbance levels both types of species can coexist – leading to a maximum in species richness.

The model by dos Santos et al. (2011) consists of 65 by 65 grid cells, each of which can be occupied by one plant species or be empty. Each species is characterised by dispersal capacity, seed production rate, adult mortality, strength of density-dependence and competitive strength. Competitive strength is defined by the chance of establishing a seedling in an empty cell where seeds from several species are present. Population dynamics of the interacting plant species are simulated in annual time steps. Disturbances occur at a certain rate and are spatially correlated. In a disturbed grid cell all individuals die.

Several variants of the model, including a neutral variant in which all species have the same characteristics, and other variants where the species characteristics differ, were analysed. Among other results, the neutral model was not able to produce a hump-shaped dependence of species richness on disturbance rate. Instead, it predicted maximal species richness at zero disturbance rate and a monotonic decline with increasing disturbance rate.

8.7 Combined Effect of Trade-Offs and Disturbance

The previously discussed hypothesis that competitively superior and disturbance-intolerant species prefer low disturbance rates while competitively inferior and disturbance-tolerant species prefer high disturbance rates has been confirmed in various model studies. Moloney (1996) modelled the impact of disturbances with different rates, timings, spatial extent, and spatial and temporal correlations on three plant species in an annual serpentine grassland located at Jasper Ridge, California. The plant species differ, among others, in their competitive strengths, dispersal abilities and time of flowering. The model simulates the life cycle of the plants in a fictitious landscape consisting of 100 by 100 grid cells of size 10 by 10 cm^2. Seed production in the model is affected by inter- and intraspecific competition, and seed dispersal is limited in distance. The analysis of the model revealed that the competitively inferior but well-dispersing species is absent at low disturbance rates while the species with high competitive strength and low dispersal ability dropped out at high disturbance rates. Coexistence of all three species was possible only under moderate disturbance rates.

The same observation that coexistence is maximised at intermediate disturbance rates was made by dos Santos et al. (2011) for four fictitious plant species with different levels of seed production, mortality, inter- and intraspecific competition and seed dispersal. In addition, the authors confirmed the relevance of the coexistence factors discussed in Section 8.1 by finding that the range of intermediate disturbance rates that allows for coexistence is widened if the strength of *intraspecific* competition is increased.

8.8 Conclusions

Understanding and predicting the dynamics of biological communities with many different species is a major challenge, and community and metacommunity models (the latter considering dispersal among communities) are important tools for addressing this challenge (Holyoak et al. 2005; Verhoef and Morin 2009; Vellend 2016; Leibold and Chase 2017). The metacommunity concept combines community ecology with the metapopulation concept. According to Leibold et al. (2004) its main paradigms include (i) patch dynamics so that habitat patches can be occupied by one or more species or not; (ii) species-sorting such that local habitat quality affects the demographic rates and thus the local abundances of the species – which thus may vary among habitat patches; (iii) mass effect, meaning that there may be non-zero net fluxes of individuals between patches, and the local extinction of a species due to environmental factors (cf. Section 6.2) or interspecific competition may be prevented through the immigration of conspecifics from other habitat patches ('rescue effect') and (iv) the neutral theory outlined in Section 8.5.

Another approach to community ecology worth mentioning but not further considered in the present book is ecological networks (Ings et al. 2009). An ecological network is defined by a set of species and their pairwise interactions. The interaction within each pair of species may be of predator–prey type ('food webs'), host–parasitoid type, mutualistic or absent (Ings et al. 2009). Macroscopic properties of the network such as its stability in the sense that no species gets displaced can be related to the network's topology, such as the number of interaction linkages in the network as well as the distribution of these linkages in the network.

This chapter introduced a number of factors that facilitate coexistence between species. The model studies presented here confirm the relevance of all factors and further indicate that the combined consideration of several or even all factors is needed to predict the biological diversity in real-world ecosystems. In addition, the combined effect of several factors may increase the range of biotic and abiotic conditions under which species can coexist.

The models that were used to investigate the biodiversity-facilitating factors encompass the full variety of ecological models presented in the previous chapters. They range from non-spatial models via patch-based metapopulation models to spatially explicit grid-based models, from deterministic to stochastic models, considering differences between different species and/or individuals to a varying degree and feedback loops as a stabilising mechanism. This again confirms that the combined use of very different model types is useful to answer complex ecological questions such as what facilitates biological diversity, and to generate new knowledge most effectively.

Part III
Economic Modelling

9 · Instruments for Biodiversity Conservation

9.1 Market Failure and the Social Dilemma of Environmental Protection

The main advantage of markets in economies is that they help allocate resources (such as land) and goods (such as corn or biodiversity) among the members of an economy efficiently so that social welfare is maximised. This also includes the decision on the efficient (or welfare-maximising) amount of a resource to be used as well as the associated efficient levels of the goods produced from the resource. Although biodiversity is often also regarded as a resource from which goods and services such as medicine and crop pollination can be produced, in the context of the present book it is considered a good that is produced at a certain cost C and has a certain benefit V to society. Since in the present context the production of biodiversity requires managing land in an appropriate manner, conserved land will often be considered a good as well.

The efficient quantity A^* of a good A is defined by the level of A that maximises the welfare function $W(A)$, which is the difference between the benefit $V(A)$ associated with A and the cost $C(A)$ of supplying A (Fig. 9.1a). The maximisation problem thus reads:

$$A^* = \arg \max_A W(A) = \arg \max_A \{V(A) - C(A)\}. \qquad (9.1)$$

If the benefit function is concave in A, that is, $\partial V/\partial A > 0$ and $\partial^2 V/\partial A^2 < 0$ (Fig. 9.1, solid lines), and the cost function $C(A)$ is convex, that is, $\partial C/\partial A > 0$ and $\partial^2 C/\partial A^2 > 0$ (Fig. 9.1, short-dashed lines), the necessary and sufficient condition for A^* is

$$\left. \frac{\partial V}{\partial A} \right|_{A=A^*} = \left. \frac{\partial C}{\partial A} \right|_{A=A^*}, \qquad (9.2)$$

that is, marginal benefit must equal marginal cost, represented by the intersection of the solid and short-dashed lines in Fig. 9.1b and leading to

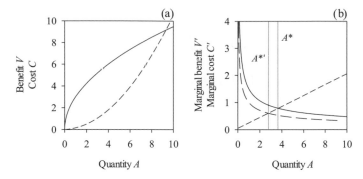

Figure 9.1 Example of a concave benefit function, $V(A) = 3A^{0.5}$ (solid line), and a convex cost function, $C(A) = 0.05A + 0.1A^2$ (short-dashed line) (panel a). Panel b shows the corresponding marginal benefit and marginal cost functions, $\partial V/\partial A = 1.5A^{-0.5}$ and $\partial C/\partial A = 0.05 + 0.2\,A$. The long-dashed line represents a reduced marginal benefit of $\partial V/\partial A = A^{-0.5}$. For the dotted lines, see text.

an efficient level of $A^* \approx 3.67$, indicated by the vertical dotted line. These conditions for the shape of the benefit and cost functions are necessary to ensure that Eq. (9.2) implies a maximum and not a minimum of welfare W.

If A denotes the amount of conserved land in a region, the assumption of a convex cost function can be explained by spatial heterogeneity in the cost of conservation and the plausible assumption that to obtain a particular level of A, the least costly land parcels are always conserved first. Thus for small A the cost per unit of conserved land is smaller than for large A, implying that $C(A)$ increases over-proportionally with increasing A. A concave benefit function can be motivated by the concept of decreasing marginal utility (cf. Section 12.1), which states that the first units of a good are valued higher than the units acquired afterwards (for instance, a bar of chocolate is usually more valuable to someone who possesses no chocolate than to someone who possesses a hundred bars of chocolate).

In the context of biodiversity conservation, a concave benefit function (i.e. a decreasing marginal benefit) can additionally be motivated if we consider the biodiversity on the conserved land as the good of interest rather than the conserved land itself. The species–area relationship (Begon et al. 1990, ch. 22.2) then states that the number of species on the conserved land increases in a concave manner with increasing size A of the conserved land, and if each additional species is valued equally each additional unit of conserved land generates less marginal benefit.

An alternative motivation for a concave benefit function is that the mean life time of a population increases in a concave manner with increasing carrying capacity if environmental stochasticity is strong (Eq. 5.18). However, there are also reasons for a convex benefit function, such as Eq. (5.18) for the case of weak environmental stochasticity, as well as thresholds effects where species can benefit from conservation efforts only if these efforts exceed a certain minimum.

If biodiversity were a private good (like a car or a chocolate bar), that is, excludable (so that the owner of some area with biodiversity could exclude others from using and enjoying it) and rivalrous (so that its use, such as one person watching and enjoying it, would diminish the possible use by another person), the marginal benefit and marginal cost functions $\partial V/\partial A$ and $\partial C/\partial A$ could be regarded as demand and supply functions $D(A)$ and $S(A)$ in a goods market. If such a goods market is perfect – in particular if property rights are well defined and information about demand (willingness of consumers to pay for the good) and supply (costs for producers to supply the good) are available to all market participants – the market delivers the efficient level of biodiversity, which is given by the equality of marginal demand and marginal supply: $D(A) = S(A)$ (Varian 2010; Hanley et al. 2007).

The problem with biodiversity, and the environment in general, from an economic point of view is that it is a public good (like other public goods such as democracy and [largely] infrastructure), which is defined by being *non*-excludable and *non*-rivalrous. To understand the consequences of the public good character of biodiversity on its provision in the society, consider two (the argument can easily be extended to $N > 2$) consumers X and Y with demand functions $D_X(A) \equiv \partial V_X/\partial A$ and $D_Y(A) \equiv \partial V_Y/\partial A$, so that each consumer wishes to maximise the difference between their benefit ($V_X(A)$ for consumer X and $V_Y(A)$ for consumer Y) and the cost $C(A)$ (Fig. 9.2).

Analogously to the discussion of Fig. 9.1, the level of biodiversity A_X^* (A_Y^*) that maximises consumer X's (Y's) difference between benefit and cost is given by the intersection of the demand curve $D_X(A)$ (respectively $D_Y(A)$) and the supply curve $S(A) = \partial C(A)/\partial A$ (Fig. 9.2, dotted lines). This implies that X and Y are willing to pay an amount, $C(A_X^*)$ and $C(A_Y^*)$ respectively, for their preferred levels of biodiversity.

Since the good A is non-excludable, if X purchases a quantity A_X^* consumer Y can use and enjoy that same quantity as well. Thus, if Y's preferred level of biodiversity is smaller than X's, that is, if $A_Y^* < A_X^*$, Y has no reason to pay any money for A and free-rides on X. Alternatively, if

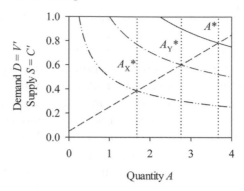

Figure 9.2 Supply (marginal cost) function $S = 0.05 + 0.1A$ (dashed line) and demand (marginal benefit) functions $D_X(A) = 0.5A^{-0.5}$ (dash–dot–dotted line), $D_Y(A) = A^{-0.5}$ (dash–dotted line) and $D_X(A) + D_Y(A) = 1.5A^{-0.5}$ (solid line). The dotted lines mark the optimal levels A_X^*, A_Y^* and A^* of A associated with the demand functions $D_X(A)$, $D_Y(A)$ and $D_X(A) + D_Y(A)$, respectively.

$A_Y^* > A_X^*$, consumer Y will spend some money but only to pay for the difference $A_Y^* - A_X^*$. Therefore, whatever the preferred levels of X and Y are, only $\max(A_X^*, A_Y^*)$ is purchased altogether (A_Y^* in the example of Fig. 9.2). Assuming that the two demand functions $D_X(A)$ and $D_Y(A)$ do not intersect (otherwise the arguments made subsequently would be slightly more complicated), the purchased level of biodiversity is given by the intersection of the supply function $S(A)$ and the largest of these two demand functions, that is, $\max\{D_X(A), D_Y(A)\}$ (which is $D_Y(A)$ in the example of Fig. 9.2) – which is the 'effective' demand for the public good.

Provision of some biodiversity level A generates a benefit $V_X(A)$ to consumer X and a benefit $V_Y(A)$ to consumer Y. Since the good A is non-rivalrous, both consumers can use and enjoy it without interfering with each other, and so the *total* benefit generated by A is the sum of the two individual benefits, $V_X(A) + V_Y(A)$. The corresponding *true joint* demand, $D_X(A) + D_Y(A)$ (solid line in Fig. 9.2), exceeds any of the two *individual* demand functions $D_X(A)$ and $D_Y(A)$. In particular, it exceeds the largest of the two demand functions, $\max(D_X(A), D_Y(A))$, that is, the demand effective in the market (dash–dotted line in Fig. 9.2). In the example of Figs 9.1 and 9.2, the efficient level of A associated with the true joint demand, $D_X(A) + D_Y(A)$, is $A^* = 3.67$ (right horizontal dotted line) which exceeds the market equilibrium level $\max\{A_X^*, A_Y^*\} = A_Y^* \approx 2.76$ (middle horizontal dotted line). Therefore, altogether, the level of bio-diversity effectively provided by the market, A_Y^*, is smaller than the efficient level A^* that is associated with the true demand and which maximises the difference between society's total benefits and costs.

This problem of underprovision is equivalent to the joint forest conservation problem discussed in Section 2.4 and typical for public goods. It represents the inability of markets to supply public goods at the efficient, that is, welfare-maximising, level – a phenomenon called market failure. Much of the environmental economics literature deals with the question of why market failure arises and how it can be fixed. The central questions are: (i) what is the welfare-maximising level of biodiversity and (ii) how can this level be reached cost-effectively, that is, at minimum economic cost (Wätzold and Schwerdtner 2005)? The present Part III, Economic Models, will deal with the second question of cost-effectiveness, while the first question will be briefly touched on in Sections 16.4–16.6.

9.2 Instruments for Biodiversity Conservation: Regulation

Perhaps the most commonly used instrument for biodiversity conservation is regulations. Most countries in the world have designated conservation areas in which the use of land is more or less restricted. The level of restriction may differ among conservation areas. In wilderness areas, for example, almost all anthropogenic activities are forbidden, even if they serve non-commercial objectives. In national parks, most types of use are restricted, so that agriculture and forestry are mostly excluded, but tourism (except in core zones) is usually allowed. Less strictly regulated conservation areas are nature parks, which usually allow extensive agriculture and forestry, but pesticides and fertilisers may be restricted or banned completely. Even outside of conservation areas agriculture is often subject to restrictions with regard to the use of chemicals or soil management, which is prescribed, for example, as 'Good Farming Practice' within the European Union's Common Agricultural Policy.

Regulations are the underlying economic concept behind most conservation plans that designate biodiversity reserves with various levels of restrictions on usage. Such plans have been developed for various regions and applications worldwide (e.g. Fernandes et al. 2005; Bode et al. 2008; Kukkala et al. 2016). Systematic conservation planning is an important field in biodiversity research (Margules and Pressey 2000; Moilanen et al. 2009), often aiming at the cost-effective selection of reserves under a budget constraint (which in the early years of the research field was considered total conserved area but now is usually measured on monetary scales).

Systematic conservation planning has been developed to make conservation decisions transparent and defensible by carrying out a number of steps, including the compilation of available data (e.g. about the spatial distribution of species and their requirements for survival as well as the biotic and abiotic parameters of [potential] reserve sites), the formulation of conservation objectives (e.g. which species to conserve and in what abundances) and the selection of conservation areas to maximise the objectives under the given budget (or to minimise the budget required to fulfil the objectives). Software tools such as Marxan (Ball et al. 2009; cf. Section 2.7) and Zonation (Lehtomäki and Moilanen 2013) have been developed for the design of cost-effective conservation plans and have been applied to many conservation problems all over the world.

The advantage of regulations is, in theory, their high level of effectiveness. If the enforcement of the regulation, including a sufficient level of monitoring and sufficiently high fines for non-compliance with the regulation, is strong enough, then the environmental target, such as the provision of a specified amount of species habitat or the reduction of chemical use, is reached practically with certainty.

A disadvantage of regulations is that in order to reach the environmental target cost-effectively, the regulator (policymaker, environmental minister, conservation agency, etc.) needs perfect information about the economic costs and the ecological benefits of all available conservation measures (including possible spatial and/or temporal variations in these quantities). The relevance of such information for the cost-effectiveness of a conservation plan has been vividly demonstrated by Ando et al. (1998) outlined in Section 3.1: if the spatial distribution of land prices is ignored and instead homogenous land prices are assumed in the selection of conservation reserves, the total cost of reaching the specified conservation target triples compared with the case in which the land prices are taken into account.

The problem is that these conservation costs are often unknown to the regulator. How costly it is to postpone the mowing of a meadow to protect breeding birds is very specifically determined, for example, by the farm structure (such as number and type of cattle and other income sources) and the famer's personal preferences regarding the environment and environmental policy, and this is difficult for the regulator to assess. Asking the farmer would in principle be possible, but of course the farmer would have a strong incentive to overstate the cost, in order to make the meadow unattractive to the regulator or to speculate on higher financial compensation. As a consequence, the desired level of conservation can be attained only at increased costs.

Since the costs of environmental measures are, as argued, often unknown to the regulator, there is no basis on which regulations can be formulated in a differentiated manner. If, for instance, the objective is to minimise the total use of agricultural chemicals in a region and it is not known how the costs of a reduced use of chemicals differ among farms or land parcels, it makes no sense to put stronger restrictions on some land parcels and weaker restrictions on others. Instead, regulations are often homogenous (there are also other political reasons for homogenous regulations, such as equity concerns; see, e.g., Pascual et al. (2010)), so that the maximum allowed level of agricultural chemicals is identical for all land parcels in a region (exceptions are, e.g., in water protection zones in which usually stricter regulations apply). An example for a homogenous obligation to provide species habitat is the Brazilian Forest Code from 1965 that demands that 20 per cent of each property is conserved for biodiversity (Chomitz 2004).

To demonstrate that homogenous regulations are not cost-effective if conservation costs are heterogeneous, consider the case of two (the calculation can be easily extended to $N > 2$) landowners (farmers, foresters or the like), $i = 1,2$, with marginal cost functions

$$c_i = 1 + e_i A_i, \tag{9.3}$$

where c_i is the cost of the last (marginal) unit of area conserved by landowner i, A_i is the total area conserved by landowner i and e_i is the slope of the marginal cost curve that tells how fast marginal cost increases with increasing conserved area. The cost c_i can involve several types of costs, such as managing costs (associated with conservation measures such as fencing), transaction costs (organisation efforts, such as acquisition of information and negotiations between regulator and landowner) and forgone agricultural profits (Naidoo et al. 2006). For simplicity, I use the term conservation costs throughout this book although in the modelling examples these usually represent forgone profits.

Marginal conservation costs increase, that is, $e > 0$, if the cost per unit area differs spatially – for example due to spatially varying agricultural productivity and hence varying profit loss associated with conservation. If a landowner decides to conserve a portion A_i of their land, they will, of course, conserve the least costly portion of the land. The more land they conserve, the more expensive is each (additional) conserved land unit.

Linearly increasing marginal costs as in Eq. (9.3) are a consequence of uniformly distributed conservation costs (Drechsler 2011; cf. Eq. (11.20)).

From Eq. (9.3) the total cost of conserving an area A_i is obtained through integration:

$$C_i(A_i) = \int_0^{A_i} c_i(A)dA = \int_0^{A_i} [1 + e_i A]dA = A_i + \frac{e_i}{2} A_i^2. \tag{9.4}$$

Let $A^{(t)}$ be the target set by the regulator for the total conserved area, $A_1 + A_2$. Under a homogenous regulation, each landowner has to conserve an area $A_i^{(t)} = A^{(t)}/2$. The total cost C_{tot} associated with conserved areas $A_1^{(t)}$ and $A_2^{(t)}$ is the sum of the integrals of the marginal costs:

$$C_{tot} = C_1\left(A^{(t)}/2\right) + C_2\left(A^{(t)}/2\right) = A^{(t)} + \frac{\left(A^{(t)}\right)^2(e_1 + e_2)}{8}. \tag{9.5}$$

In contrast, the cost-effective sizes of the conserved areas, A_1^* and A_2^*, are obtained by solving the optimisation problem

$$\min C_{tot} \quad \text{s.t.} \quad A_1 + A_2 \geq A^{(t)}$$
$$C_{tot} = \int_0^{A_1*} c_1(A_1)dA_1 + \int_0^{A_2*} c_2(A_2)dA_2. \tag{9.6}$$

Forming the Lagrangian

$$L = C_{tot} - \lambda\left(A_1 + A_2 - A^{(t)}\right) \tag{9.7}$$

and setting its derivatives with respect to A_1, A_2 and λ to zero delivers the optimality condition

$$c_1(A_1) = \frac{dC_{tot}}{dA_1} = \lambda = \frac{dC_{tot}}{dA_2} = c_2(A_2) \tag{9.8}$$

and the area constraint $A_1 + A_2 = A^{(t)}$. The optimality condition tells us that total cost C_{tot} is minimised if the marginal costs $c_1(A_1)$ and $c_2(A_2)$ are equal. Inserting Eq. (9.3) together with the constraint $A_1 + A_2 = A^{(t)}$ into Eq. (9.8) yields the cost-effective levels of A_1 and A_2:

$$A_1^* = \frac{A^{(t)}}{1 + e_1/e_2} \quad \text{and} \quad A_2^* = \frac{A^{(t)}}{1 + e_2/e_1}. \tag{9.9}$$

Equation (9.9) implies that if the two marginal cost functions are equal, $e_1 = e_2$, both conserved areas should be equal, while in the case of unequal marginal cost functions the landowner with the higher cost slope e should conserve less than the one with the smaller cost slope. The total cost associated with this cost-effective allocation is

$$C_{tot}^* = A^{(t)} + \frac{\left(A^{(t)}\right)^2}{2}\left(\frac{e_1}{(1 + e_1/e_2)^2} + \frac{e_2}{(1 + e_2/e_1)^2}\right). \tag{9.10}$$

This is smaller than the total cost C_{tot} under homogenous regulation (Eq. 9.5) and the efficiency loss associated with homogenous regulation, that is, the difference between C_{tot} of Eq. (9.5) and C_{tot} of Eq. (9.10), equals

$$\Delta C_{tot} = \frac{\left(A^{(t)}\right)^2}{4}\left(\frac{e_1 + e_2}{2} - \frac{2e_1e_2}{e_1 + e_2}\right). \tag{9.11}$$

The first of the two fractions in the parentheses represents the arithmetic mean and the second fraction represents the harmonic mean of e_1 and e_2. If the marginal cost functions are identical, $e_1 = e_2$, the two means are equal, the efficiency loss is zero and the homogenous allocation is cost-effective. However, if e_1 and e_2 differ, the arithmetic mean exceeds the harmonic mean, the efficiency loss is positive and the homogenous allocation is not cost-effective.

9.3 Instruments for Biodiversity Conservation: Conservation Payments

As previously demonstrated, regulations are likely to be inefficient if the conservation costs vary among landowners and are not known to the regulator. Another disadvantage of regulations is that they are not voluntary, so landowners have to obey them and are fined if they resist doing so. This may erode the acceptance of biodiversity conservation in the community of landowners.

Here market-based conservation instruments are a valuable alternative. They promise a higher level of cost–effectiveness and are voluntary. The most popular market-based biodiversity conservation instrument world-wide is conservation payments (Ferraro and Simpson 2002; Engel et al. 2008). In a conservation payment scheme, each landowner is offered a payment p (usually per unit of land) if they manage the land in a biodiversity-friendly manner, that is, they apply certain conservation measures or avoid harmful land-use practices. Each landowner can decide either to manage the land in a biodiversity-friendly manner and receive the payment, or to manage it otherwise (most likely in a manner that maximises agricultural or other profits).

Conservation payments are applied practically on all continents of the world. In Europe, for example, they are known as agri–environmental schemes (AES) (Bátary et al. 2015). One of the oldest examples is the meadow bird programme that is available, for example, to farmers in most federal states of Germany. Many meadow birds breed in spring, which is usually also the time of the first harvest. If a meadow is mowed before the juvenile birds can fly, all or most of the offspring are killed. To protect these birds, farmers are paid if they postpone the cut of the meadow. In Bavaria, for example, farmers receive €230/ha or €320/ha if they do not mow their meadow before 1 June or 15 June, respectively (VNP 2018).

Most AES are homogenous in the sense that the same payment is offered to all farmers – for the same reasons regulations are usually homogenous: setting payments individually for each farm or land parcel would require knowledge about the farm-specific (or parcel-specific) conservation costs, and unequal payments for one and the same conservation measure could raise equity concerns (see, e.g., Pascual et al. (2010)). Nevertheless, homogenous payment schemes are, in contrast to regulations, cost-effective even in the case of heterogeneous conservation costs, as the following calculations will show.

Assume again two landowners, $i = 1,2$ (an extension to $N > 2$ is straightforward) with marginal cost functions given by Eq. (9.3). If a payment p is offered for each unit of conserved area the total profit of landowner i who conserves an area A_i is

$$\Pi_i = pA_i - C_i(A_i). \tag{9.12}$$

The area A_i^* that maximises profit Π_i is obtained by differentiating Eq. (9.12) with respect to A_i and setting the obtained marginal profit

$$\pi_i = \frac{d\Pi_i}{dA_i} = p - c_i(A_i) \tag{9.13}$$

to zero. The profit-maximising level A_i^* of conserved area thus is where the payment p equals the marginal cost $c_i(A_i^*)$. The economic reason is that if less area than A_i^* was conserved, the conservation of an additional unit of land would cost less than would be gained from the payment p, while if more than A_i^* was conserved the conservation of an additional unit of land would incur higher costs than would be gained by the payment.

Altogether, each landowner chooses their conserved area so that the marginal cost equals the payment, which implies that the payment induces equality of the marginal costs of both landowners. Since equality of marginal costs implies cost-effectiveness (Eq. 9.8), payment schemes are 'by definition' cost-effective. This is regardless of the level of the payment p which only determines the total conserved area $A^{(t)}$ and the associated total conservation cost.

A final consideration shall take up the issue of the provision of public goods raised in Section 9.1. For this, assume N landowners, $i = 1, \ldots, N$, in some region. Each landowner has the choice of conserving a certain area a of their land or using it for economic purposes. The cost of conserving area a incurs a cost c_i that differs among landowners. If the c_i are ordered from smallest to largest and plotted from left to right, one obtains a 'regional' marginal cost curve (not to be confused with the marginal cost curves of the individual landowners above) that shows the conservation cost c_i per unit area a as a function of total conserved area A (Fig. 9.3, solid line).

If a payment p is offered for conservation (Fig. 9.3, dashed line), all landowners with cost $c_i < p$ will conserve and the others will use their land for economic purposes. In Fig. 9.3, the conserving landowners are those with areas to the left of the intersection of the solid and dashed lines. By comparing Figs 9.1b and 9.3, one can identify the solid regional marginal cost curve of Fig. 9.3 with the supply function of Fig. 9.1b, and

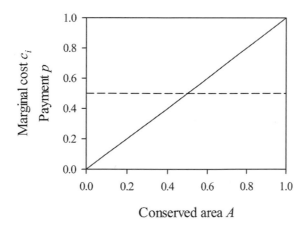

Figure 9.3 Regional marginal cost c_i, scaled between zero and one (solid line), and payment p (dashed line) as functions of conserved area A, scaled between zero and one.

the dashed line in Fig. 9.3 with the demand curve of Fig. 9.1b. The only difference is that in Fig. 9.3 the demand curve is flat, meaning infinite demand for areas with low conservation costs $c_i < p$ and zero demand for areas with high conservation costs $c_i > p$. The market equilibrium is obtained where the regional marginal cost c_i equals the payment p – with equilibrium conserved area A^* (equalling 0.5 in the example of Fig. 9.3).

This analysis allows us to formulate three conclusions. Firstly, since the landowners with the lowest costs conserve their land, the allocation induced by the payment scheme is cost-effective in that the targeted amount of area is conserved at least cost.

Secondly, a conservation payment scheme can 'fix' the problem of market failure that arises from an underrated demand for biodiversity by establishing an artificial demand function (dashed line in Fig. 9.3), or in other words by assigning biodiversity (in the present analysis: conserved area) a market price. This market price serves as a signal to landowners to supply biodiversity (conserved area) in the demanded quantity A^*. If p is chosen according to society's true demand for biodiversity, the payment scheme delivers the efficient level of conserved area.

This construction is equivalent to the so-called Pigouvian tax, named after economist Arthur Cecil Pigou (1877–1959). If a (Pigouvian) tax is put on environmentally harmful behaviour, a rationally operated firm will avoid that behaviour to a level at which its marginal avoidance cost equals the tax (the equivalent to the equation $p = c_i(A_i)$ derived from Eq. (9.13)). If the Pigouvian tax is set equal to society's marginal damage cost incurred by the environmental degradation (the equivalent to the dashed line in Fig. 9.3), the environment is degraded exactly to the level that maximises social welfare, that is, the level at which the sum of environmental damages and avoidance costs is minimal. By this construction the external costs of environmental degradation are internalised in the market, so they are accounted for by the economic actors.

The third conclusion from Fig. 9.3 is that by setting the payment at a certain level the regulator sets the price p of conserved area, and the supplied quantity of area, A^*, emerges from the decisions of the profit-maximising landowners. Therefore payments, as well as taxes, belong to the class of so-called price-based policy instruments.

9.4 Instruments for Biodiversity Conservation: Offsets

An alternative to conservation payments is tradable permits, also termed conservation offsets (Tietenberg 2006; Hanley et al. 2007, ch. 5). In a conservation offset scheme each landowner who conserves land receives credits, usually proportional to the amount of conserved area A_i (assuming such proportionality, I will denote both conserved area and associated credits by the same variable A_i). To initiate conservation activities, the regulator sets for each landowner i a target $A^{(t)}$ for conserved area. Landowners who provide more conserved area than the target $A^{(t)}$ can sell the associated credits $A_i - A^{(t)}$ on the market. Other landowners who wish to conserve less than the target can do so, but they must buy a corresponding amount of credits on the market. Naturally, landowners with low conservation costs will tend to conserve more than required and sell credits while landowners with high conservation costs will tend to conserve less and buy credits (Wissel and Wätzold 2010).

To model this process, consider again two landowners, $i = 1,2$ (an extension to $N > 2$ is straightforward), with cost functions given by Eq. (9.3). The profit of each landowner is composed of the revenues from selling credits at some price $\hat{p}$ and the conservation cost:

$$\Pi_i = \hat{p}(A_i - A^{(t)}) - C_i(A_i). \qquad (9.14)$$

Assuming each landowner is a price taker, that is, ignoring issues of bargaining power, monopolistic behaviour and so on (e.g. Varian 2010), each landowner chooses, for given credit price $\hat{p}$, the profit-maximising level of A_i. Analogously to Section 9.3, this level is obtained by differentiating Eq. (9.14) with respect to A_i and setting the resulting marginal profit

$$\pi_i = \frac{d\Pi_i}{dA_i} = \hat{p} - c_i(A_i) \qquad (9.15)$$

to zero. In market equilibrium both landowners experience the same credit price $\hat{p}$, so $\pi_i = 0$ implies that the marginal costs c_i of both landowners are equal. Similar to Section 9.3, together with Eq. (9.8) one can conclude that in market equilibrium the conservation efforts A_i are allocated cost-effectively between the two landowners, so that the conservation target $2A^{(t)}$ is reached at minimum total cost $C_1 + C_2$.

To obtain the associated conserved areas A_1^* and A_2^*, the equation $c_1(A_1^*) = c_2(A_2^*)$ must be solved together with the constraint that in market equilibrium supply of credits equals demand:

$$(A_1 - A^{(t)}) + (A_2 - A^{(t)}) = 0. \qquad (9.16)$$

This equation simply renders the regulator's regional conservation target that $A_1 + A_2 = 2A^{(t)}$.

Readers who are acquainted with climate policy may have immediately noticed that the outlined conservation offset scheme is almost identical to the emissions trading schemes that have been established, for example, for CO_2 emissions control in Europe and a number of states in the United States. The main difference is simply that in emissions trading a cap is set that must not be *exceeded* (and reducing emissions towards the cap is costly) while in conservation offsets a target must not be *undercut* (and it is costly to increase conservation efforts towards the cap). Tradeable permit schemes are increasingly used in various fields of environmental policy (Tietenberg 2006; Hansjürgens et al. 2016). In the context of biodiversity conservation they are usually termed conservation offsets and are applied, for example, as conservation banking in the United States (Fox and Nino-Murcia 2005) and eco-accounts in Germany (Wende et al. 2018). For Brazil, the potential efficiency gains from a conservation offset scheme for forest protection have been calculated by Chomitz (2004).

Similar to the previous section, the functioning of a conservation offset scheme can be related to the public good problem of Section 9.1. For this, again assume a region with N landowners, each of which can choose between managing some area a for conservation or for economic purposes. Conservation costs c_i again differ among landowners and can be arranged into a regional marginal cost curve (Fig. 9.4, solid line).

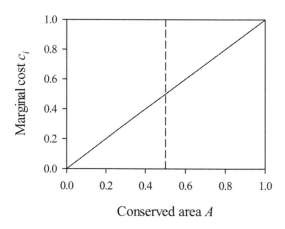

Figure 9.4 Regional marginal cost c_i, scaled between zero and one (solid line) as a function of conserved area A, scaled between zero and one. The dashed line represents the area target $NA^{(t)}$.

If a target is set by the regulator such that each landowner has to conserve at least an area $A^{(t)}$, then the total area that needs to be conserved equals $NA^{(t)}$. This constraint again can be regarded as a demand curve, which is now a vertical line (Fig. 9.4, dashed line). This means that if the amount of conserved area is below the target $NA^{(t)}$ the willingness to pay for an additional conserved area is infinite, and if the amount of conserved area is above the target the willingness to pay is zero. Since landowners are allowed to trade, the least costly areas to the left of the dashed line are conserved (and credits sold on the market) while the owners of the other areas buy credits and use their land for economic purposes. The marginal cost associated with the intersection of the solid cost curve and the dashed target line in Fig. 9.4 is identified as the equilibrium credit price.

Analogously to Section 9.3, a first conclusion is that the conservation offset scheme induces the cost–effective allocation of conservation efforts. Secondly, Fig. 9.4 visualises that in conservation offsets the problem of an underrated demand for biodiversity is fixed by establishing an artificial demand function in the form of the target $NA^{(t)}$ for conserved area (biodiversity). The credit price again serves as a market signal for landowners to provide conserved area. If the target $NA^{(t)}$ is chosen according to society's true demand for biodiversity, the conservation offset scheme delivers the efficient level of conservation.

Lastly, in contrast to the conservation payments of Section 9.3, in conservation offsets the regulator sets the quantity $NA^{(t)}$ of the environmental good to be supplied, and the price emerges from the transactions among the landowners. Therefore, tradable permit schemes in general and conservation offsets in particular belong to the class of so-called quantity-based policy instruments.

9.5 Conclusions

As has been shown, market-based instruments such as payment schemes and tradable permit schemes promise a higher level of cost-effectiveness than regulations. However, in the application of market-based instruments to the conservation of biodiversity one should also be aware of some challenges and limitations.

A general problem related to market-based instruments is uncertainty in the supply curve. In conservation payments the supplied quantity for a given payment level depends on the shape of the supply curve, that is, the marginal costs of providing the good. If this is not known, the quantity induced by a certain payment level is subject to uncertainty – and so is the budget that is required to finance the provision of the good, because it is given by the product of payment and quantity. Although the payment scheme is cost-effective in that the quantity is maximised for given costs, the effectiveness, that is, the level of the quantity provided, may be uncertain (Hanley et al. 2007, ch. 5.6.2). On the other hand, in a tradable permit scheme the total cost associated with a given target level of the quantity depends on the supply curve, and if that supply curve is uncertain so is the total cost (Hanley et al. 2007, ch. 5.6.2) – which may have adverse consequences for the economy.

If payment schemes or tradable permit schemes are applied to bio-diversity, additional problems may arise that are due to the particular nature of biodiversity: biodiversity is heterogeneous in type (i.e. species, populations, individuals: cf. Chapter 7), spatial structure (e.g. the viability of a population often depends on the spatial arrangement of its habitat: cf. Chapter 6) and time (i.e. ecosystems evolve over time). All of these issues provide limitations to market-based instruments that usually assume that the quantity to be produced is homogenous, so that one unit of the good is identical to any other, and it makes no difference when and where this unit is provided. An example of such a homogeneous good in the field of environmental protection is CO_2: in its impact on the climate a ton of CO_2 is a ton of CO_2 wherever it is emitted and, to a certain extent, whenever it is emitted.

If, in contrast, a market-based instrument is applied to the protection or provision of a heterogeneous good such as biodiversity, we face various trade-offs (Wissel and Wätzold 2010; van Teeffelen et al. 2014). If we ignore the heterogeneity and, for instance, treat a hectare of natural forest as a hectare of extensive grassland, or a hectare of beech forest in the German state of Thuringia as a hectare of beech forest in the neighbouring state of Hessia, then we can readily apply the market-based instruments to protect the goods' 'natural habitat' or 'beech forest', respectively. But then we can never be sure that in time we will not lose all natural forests and end up with only extensive grassland, or lose all beech forest in Thuringia and end up with beech forest only in Hessia.

To avoid these uncertainties in the conservation outcome, the instruments must be designed in a more targeted manner and the environmental good defined more specifically. This, however, comes only at a cost. For instance, a payment scheme would have to be established for each type (natural forest and extensive grassland, respectively), which increases the transaction costs of setting up and running the schemes. If the creation of a habitat close to other habitats is more effective for the survival of a target species than the creation of a spatially isolated habitat, then the former action should be rewarded a higher payment than the latter – again increasing transaction costs.

If the environmental good traded in a permit scheme is defined too narrowly (e.g. the presence of a particular rare tree species in a particular forest), only a few landowners will be able to supply that good and the number of market participants will be too small to allow for a functioning and efficient market (Wissel and Wätzold 2010). Likewise, similar to the case of the conservation payments, if the location of a created habitat relative to other habitat matters, the trading rules must be formulated in a more complicated manner – increasing transaction costs.

Next to these biophysical limitations to market-based biodiversity conservation instruments, additional obstacles (some of which, however, can also occur in regulatory policies) include weakly defined property rights (e.g. ownership of land is not clearly specified), lack of enforcement power (e.g. the regulator cannot sanction a landowner who receives a payment but refuses to carry out the associated conservation measure), lack of additionality (landowners receive payments for measures they would have taken anyway), leakage (conservation in one region increases development pressure in other regions) and others (Engel et al. 2008; Walker et al. 2009).

Altogether, market-based instruments have a high potential to conserve biodiversity more cost-effectively, but they are no cure-all for the problems of biodiversity loss, and their pros and cons must be weighted carefully when applying them to particular cases. Nevertheless, for biodiversity conservation on private lands such as agricultural landscapes they are an indispensable alternative to regulations, and are increasingly used on all continents (Schomers and Matzdorf 2013).

Recent literature related to Chapter 9 includes Ring et al. (2010), Chobotová (2013), Santos et al. (2014), Rode et al. (2016) and Schirpke et al. (2018).

10 · Game Theory

10.1 Public Good Problems, External Effects and the Prisoner's Dilemma

In Section 9.1 we saw that public goods such as biodiversity are often undersupplied because people free-ride by consuming the good without paying (enough) for it. A simple, practical example of such a public good problem was presented in Section 2.4 where two countries could decide between conserving and developing their share of a cross-border forest. It was argued that both countries would be better off by both conserving, but since each country itself is better off by developing, regardless of the decision of the other country, both countries end up developing – and thus reach a state in which both are worse off.

In game theory (e.g. Fudenberg and Tirole 1991; von Neumann and Morgenstern 2007; Tadelis 2013) such a dilemma is called a prisoner's dilemma (Kollock 1998). The term prisoner's dilemma stems from the fictitious case of two criminals who could minimise their *joint* sentence by cooperating with each other and being silent in front of the prosecutor, but could minimise their *own individual* sentence by defecting and betraying the other suspect. As in the example of conserving a joint forest, cooperation would maximise the joint pay-off, but both suspects will eventually defect and end up in a state in which they are both worse off.

The underlying cause of public good problems is the presence of external effects. In the case of the forest, conservation of the forest by one country creates a positive externality for the other country, such that the other country's pay-off is increased (in Table 10.1 by $80 million). Conversely, if a country develops its forest it creates a negative externality for the other country by reducing that country's pay-off.

Environmental externalities are often of a spatial nature such that an action on one site affects the conditions and the impacts of actions on other sites. An instructive example of this is the protection of pollinators

Table 10.1 *Profit of landowner A after Eq. (10.2) as a function of each landowner's choice to spray or conserve.*

		Landowner B	
		Conserve	Spray
Landowner A	Conserve	0.8	−0.1
	Spray	0.9	0

in agricultural systems whose presence has significant effects on agricultural returns (Gallai et al. 2009). Landowners in a region typically have the choice between spraying their land to kill pests or refraining from spraying. Choosing the latter will likely reduce harvest and agricultural returns, but it may also conserve pollinators that would be killed by spraying. The presence of the pollinators in turn increases the local returns on the conserved land and in addition also increases the returns on the neighbouring land if the pollinators are mobile and pollinate the plants on neighbouring land as well. Therefore pollinator conservation causes local costs (which call for 'defecting' and spraying) but generates positive externalities for neighbouring landowners.

This dilemma can be formalised (Drechsler 2018) by modelling the profit of an isolated land parcel i as

$$\pi_i = (\beta - c)x_i, \tag{10.1}$$

where $x_i = 0$ if the land parcel is sprayed and $x_i = 1$ if the land is not sprayed and the pollinators are conserved. If the land parcel is sprayed, the profit (without loss of generality) is set to zero. If the land parcel is conserved, pest species reduce the profit by some magnitude c while pollinators increase it by some magnitude β.

If the land parcel is surrounded by other land parcels that contain pollinators, the pollination benefit increases to βn_i where n_i is the total number of conserved land parcels in the neighbourhood (including parcel i), and the profit of land parcel i is

$$\pi_i = \beta n_i - c x_i \tag{10.2}$$

(note that Eq. (10.1) is a special case of Eq. (10.2), since for an isolated land parcel $n_i = x_i$). Now consider the case of two adjacent landowners, with β being a little smaller than c, for example $c = 1$ and $\beta = 0.9$ (if $\beta > c$ there is no dilemma and both landowners maximise their individual profits by conserving, $x_i = 1$). The corresponding profit π_A for landowner A is given in Table 10.1.

If both landowners cooperate and conserve, the profit of each landowner is 0.8, which exceeds the profit of zero that would be obtained if both landowners defected and sprayed. However, from the point of view of landowner A's profit alone, spraying always pays and increases landowner A's profit by 0.1. Since the game is symmetric, the analogue holds for landowner B, and both landowners maximise their individual profit by spraying – even though this decision reduces the profit of each landowner by 0.8.

Spraying maximises the individual profit independent of the land-use decision of the other landowner, so spraying (defecting) is a so-called dominant action. (The more common term is 'dominant strategy'. However, I will use the term 'strategy' in Section 10.3 to denote a rule set that tells whether to cooperate or defect as a function of the circumstances. To avoid confusion, I save the term 'strategy' for such a rule set and use 'action' to denote whether a player actually cooperates or defects.)

The same results are obtained for other pay-offs, as long as they fulfil certain constraints. If one denotes $\Pi(x_A, x_B)$ the pay-off for player A (the same can be done for player B) where $x_A \in \{C,D\}$ and $x_B \in \{C,D\}$ are the chosen actions of player A and player B, and C and D represent cooperation and defection, respectively, the following inequalities must hold for a prisoner's dilemma:

$$\Pi(D, C) > \Pi(C, C) > \Pi(D, D) > \Pi(C, D), \qquad (10.3)$$

that is, player A fares best if they defect while player B cooperates, and worst if they cooperate while player B defects; the pay-offs for the cases of both players cooperating and both players defecting lie in between.

10.2 A Few Bites of Game Theory

In Section 10.1 we saw that the public good problem can be analysed very intuitively within the framework of game theory. Game theory is at the intersection of economics and mathematics and was the first theory analysing and understanding the behaviour and actions of cognitive individuals such as humans and many animals. It represents one of the foundations of modern economics and is, among others, an important approach to the design and analysis of economic incentives (Chapter 11).

Table 10.2 *Example of a coordination game: pay-off of player A as a function of each player's choice to cooperate or to defect.*

		Player B	
		Cooperate	Defect
Player A	Cooperate	0.8	−0.1
	Defect	−0.1	0

The following two sections outline a few basics of game theory, which of course can encompass only a tiny fraction of game theory. Numerous textbooks have been written on the topic, such as Shoham and Leyton-Brown (2008), Gintis (2009) and Tadelis (2013).

Many games have the same general structure as the prisoner's dilemma and can be generated by varying the pay-offs in Table 10.1 (Wiki 2018). Assuming that mutual cooperation always leads to higher pay-offs than mutual defection, and considering that adding a constant to all pay-offs does not change the type of game, the pay-offs $\Pi(C,C)$ and $\Pi(D,D)$ (with $\Pi(C,C) > \Pi(D,D)$) can be kept fixed without loss of generality, and only the pay-offs $\Pi(C,D)$ and $\Pi(D,C)$ need to be varied.

A popular relative of the prisoner's dilemma is the coordination game obtained by assuming that $\Pi(C,D) < \Pi(C,C)$ and $\Pi(D,C) < \Pi(D,D)$. From the example of Table 10.1, this can be obtained, for example, by reducing the pay-off of player A for the case in which they defect while player B cooperates from 0.9 to −0.1(Table 10.2).

In this case defecting is no longer a dominant action, since the best response to player B cooperating is to cooperate as well, while defection maximises the pay-off only if player B defects too. As a consequence, mutual cooperation, which was not a 'rational' outcome in the prisoner's dilemma, now becomes rational. More precisely, both mutual cooperation and mutual defection are so-called Nash equilibria (Fudenberg and Tirole 1991; von Neumann and Morgenstern 2007; Tadelis 2013). In a Nash equilibrium none of the players have an incentive to change their actions from the equilibrium one. This does not mean that the currently chosen action maximises the player's profit *regardless* of the other player's actions, but since in the Nash equilibrium none of the other player(s) have an incentive to change their action, each player can expect the other(s) to stay with their current action. Thus, nobody will change their actions and the Nash equilibrium is stable.

Table 10.3 *Example of a chicken game: pay-off of player A as a function of each player's choice to cooperate or to defect.*

		Player B	
		Cooperate	Defect
Player A	Cooperate	0.8	0.1
	Defect	0.9	0

Faced with a prisoner's dilemma, in the state of mutual defection neither of the two players has an incentive to change their action, since cooperation will reduce the pay-off, and even in the state of mutual cooperation it pays to defect. The only Nash equilibrium in the prisoner's dilemma therefore is the state in which both players defect. In the coordination game of Table 10.2, in contrast, the pay-off is reduced if the state of mutual cooperation is left by defecting, so both the states of mutual cooperation and mutual defection are Nash equilibria. As a corollary, dominant actions are always stable in the sense of a Nash equilibrium, but a Nash equilibrium is not necessarily composed of dominant actions.

Another game that is as relevant as the prisoner's dilemma is the chicken game, or hawk–dove game, with pay-offs

$$\Pi(D, C) > \Pi(C, C) > \Pi(C, D) > \Pi(D, D). \tag{10.4}$$

Compared with Eq. (10.3), the third inequality has been reversed, so if player B defects it is better for player A to cooperate. A numerical example is shown in Table 10.3 in which only the sign of the pay-off $\Pi(C,D)$ has been switched compared with Table 10.1. One interpretation of these inequalities is two car drivers approaching each other from the opposite ends of a narrow road. The driver who gives way first may feel like a coward, or a 'chicken', so the pay-off of cooperation is low, as in the prisoner's dilemma. However, in contrast to the prisoner's dilemma, if both drivers defect and refuse to give way the pay-off of both players is minimised (whatever that means in practice).

The chicken game has many applications in real-world conflicts. Many negotiations in the socioeconomic arena are of a chicken-type. If, for instance, a labour union wants to obtain a wage rise for its members it may threaten uncooperative employers with a strike. Such a strike would be costly for both sides, since the employers could not produce and the employees would lose wages for the duration of the strike (more drastic consequences could even be thought of for both sides). Therefore there is some incentive for both sides to be chicken and back down: if the employer raises wages as demanded by the union the strike is averted, but this comes at the cost of higher labour costs; if the union backs down the strike is averted, too, but this comes at the opportunity cost of a lost wage rise; and in addition, whoever backs down will lose reputation for future negotiations. Such conflicts are usually solved through a compromise that allows both sides to save face. Considering such options formally would require extending the discrete decision space of the present game to a continuous one, which however would be beyond the scope of this brief outline.

In ecology, the chicken game is known as the hawk–dove game and is used to model, for example, conflicts between individuals over territories or resources. Two individuals playing 'dove' will share the contested resource, and when in conflict with an individual choosing 'hawk' a 'dove' player will back down and waive its share of the resource. Consequently, when in conflict with a 'dove' player it pays to play 'hawk' and pocket the entire resource. However, if two 'hawks' encounter each other they will escalate their conflict, possibly leading to high costs through injury or death – which exceed the expected benefit from the resource. Consequently, when in conflict with a 'hawk' it pays to be a 'dove'.

These examples of games belong to the class of non-cooperative, symmetric, non-zero-sum, static, simultaneous games. To explain these terms one by one, in a non-cooperative game, agreements must be self-enforcing, which means that cooperation between individuals is stable only if it is a Nash equilibrium. In contrast to this, cooperative game theory studies situations in which agreements are externally enforced, and typical questions here are what kinds of agreements are established and what their joint pay-offs are.

In a symmetric game the pay-offs only depend on the actions chosen, not on who chooses them, so the identity of the players can be exchanged without changing the outcome of the game. This is different in asymmetric games such as the ultimatum game (see Section 12.3) in which one player can decide how to share a common resource and the other can decide whether to agree to that share or reject the suggestion. In the case of rejection no one gets the resource.

A non-zero-sum game models situations in which the joint pay-off, that is, the sum of the pay-offs for all players, is not constant. The clue in the prisoner's dilemma, for example, is that the joint pay-off is larger if both players cooperate than if they both defect. Chess, in contrast, is an example of a zero-sum game in which the gain of one player equals the loss of the other.

In a static game players encounter each other only once, while in a dynamic game they meet several times (cf. Section 10.3). In simultaneous − static or dynamic − games all players decide on their actions at once, without knowing what the other players do. In contrast, in sequential games one player moves first, followed by the second player and so on. An example of a sequential game from the field of biodiversity conservation is by Bode et al. (2011), who investigate the outcomes of different land acquisition strategies of two conservation organisations.

10.3 Dynamic Games and Evolutionary Game Theory

In dynamic games the players meet several times and the circumstances under which they meet may change. In such settings the question arises, under which circumstances does a player choose to cooperate and under which to defect? I term the rule set that specifies under which circumstances to cooperate and when to defect a strategy. Among the many strategies that can be thought of, the simplest ones are to cooperate or defect unconditionally, which may be denoted as AllC and AllD, respectively. Another frequently considered strategy is tit-for-tat (TFT) where a player cooperates if the other player(s) cooperated in the previous encounter, and defects otherwise. Many variants of this strategy exist, such as generous tit-for-tat (GTFT) where the player cooperates even after the other player(s) has defected once, and only after the other player(s) have defected a second time does the GTFT player eventually defect too. Both these and other variants of TFT can be extended by a stochastic element, so that with a certain small probability the chosen action differs from what the deterministic rules say.

TFT is a very interesting and relevant strategy both from an environmental-economic and a behavioural science point of view. On the one hand it includes cooperation and has the potential to allow for escape from the deadlock of the inefficient Nash equilibrium in the prisoner's dilemma with all players defecting. On the other hand, since it responds to defection with defection it is 'immune' to exploitation by defectors who in the prisoner's dilemma benefit from the cooperation of others at the cost of these cooperators. Axelrod and Hamilton (1981) simulated a 'tournament' of strategies, including AllC, AllD, TFT and a number of others, that mimicked an evolutionary process where in an encounter between a 'superior' and an 'inferior' strategy (player) the superior one produced more offspring than the inferior one, and where superiority was measured by the pay-off ('fitness') of the strategy (player) in the encounter. In the long run of the tournament, TFT was the most successful strategy.

Since it responds to cooperation with cooperation and to defection with defection, TFT can be regarded as a simple model of reciprocal altruism in which an individual responds to altruism with altruism and, for example, shares food with another individual (or pays for a round in the pub) if that individual did the same for them in the past, or at least can be expected to do so in the future (Nowak 2006).

The evolution of strategies is the core theme of evolutionary game theory (Maynard Smith 1982; Weibull 1995). Evolutionary game theory couples classical game theory with processes of Darwinian evolution, including mutation and selection (Vincent 2005). The game pay-offs serve as means to assess the relative advantage of one strategy over another. In an encounter of two strategies, the strategy with the higher pay-off will replace the inferior strategy (quantitatively modelled by a so-called replicator equation), so in the long run its abundance in the population is likely to increase while that of the other strategy is likely to decline.

Two questions of particular interest in evolutionary game theory are whether a particular strategy survives in the population, that is, it is played by at least one individual even after a large number of simulation steps, and whether a particular strategy can invade the population, that is, it is played by a significant proportion of individuals even if it was initially played only by a single individual.

Whether a behavioural trait such as equity preference and altruism or a strategy such as AllC or TFT survives in the presence of AllD depends to a large extent on the interaction between the players. This includes, among others, whether a player can potentially encounter *any* other player in the population or only players within a certain (spatial) neighbourhood.

An early paper considering the relevance of spatial structure is Nowak and May (1992), who considered a square grid of cells, each of which hosts a player who can choose between the two strategies AllC and AllD. In each time step the player may or may not change their strategy. The player does this by comparing their pay-off with that of the eight neighbours in the Moore neighbourhood and choosing the strategy associated with the highest pay-off.

In the model simulation of Nowak and May (1992) each player played a prisoner's dilemma against each of their eight neighbours, the pay-off for each of the eight encounters was recorded and a sum over all eight pay-offs was built. After this each player compared this summed pay-off with the summed pay-offs of the eight neighbours (who had also played against their eight neighbours) and for the next time step chose the strategy of the player (themself and the eight neighbours) with the highest summed pay-off.

In the simulation it turned out that cooperation (AllC) can coexist with defection (AllD). This is possible because it turns out that the cooperators (as well as the defectors) aggregate in space, so a cooperator is surrounded mainly by cooperators and a defector mainly by defectors. Given the nature of the prisoner's dilemma, a cooperator surrounded by cooperators always fares better than a defector surrounded by defectors, so by the rules of Nowak and May's game the cooperators do not switch to defection (depending on the pay-offs in the pay-off matrix, there may be some turnover between the two strategies, but the general explanation is the same), and so cooperation survives as a strategy in the population.

A similar analysis was done by Brauchli et al. (1999), who considered two additional strategies: TFT and Pavlov, where Pavlov means that a player stays with the previous action (cooperate or defect) if that led to the highest pay-off, and switches to the other strategy otherwise ('win stay – lose shift'). The encounter with the eight neighbours was modelled in the same way as in Nowak and May (1992), but in addition, to model the strategies TFT and Pavlov, each player played each of their eight neighbours for a number of 'fictitious' rounds, and the pay-offs of these fictitious rounds were summed up.

A main result of this model analysis is that TFT can survive and is not fully replaced by AllD or any other strategy, which confirms that reciprocal altruism is a viable concept to overcome the socially sub-optimal Nash equilibrium of mutual defection. The question arises, however, how realistic the chosen replicator model is that determines under which condition a player changes their strategy. Firstly, one may wonder if a player (landowner) in the real world has the chance to play against each of their neighbours many times to assess the summed pay-off of their own strategy, and what the practical meaning of that summed pay-off is. Secondly, although a cooperator surrounded by cooperators fares better than a defector surrounded by defectors (largely driving the results of Nowak and May 1992), a player surrounded by cooperators still fares better if they defect (Table 10.1). Therefore one may wonder why a cooperator (if they are truly self-interested) should not switch to defection and exploit their cooperating neighbours?

An alternative replicator model may be for a focal player to randomly select an opponent from among their eight neighbours. Both opponents now choose to cooperate or defect – depending on their current strategy (AllC, AllD, TFT, etc.) and the previous actions (cooperate or defect) of their respective opponent. If the resulting pay-off of the opponent exceeds the pay-off of the focal player, the focal player adopts the opponent's strategy, but otherwise sticks to their own strategy.

Drechsler (2018) simulated this evolutionary game and found that TFT still survives even in the presence of AllD. In particular, at the end of the simulation about 50 per cent of all players were playing AllD and about 50 per cent were playing TFT. A closer look at the chosen *actions* (cooperation or defection), however, revealed that all the TFT players actually defected – in response to the defection of the AllD players. Only if the pay-offs in the pay-off matrix were changed (effectively considering games other than the prisoner's dilemma) so that AllC became a viable strategy, too, did some TFT players actually cooperate – in response to the cooperation of the AllC players. This result highlights that the altruism captured in the TFT strategy is really only of a reciprocal nature and a TFT player effectively cooperates only if there are unconditional cooperators ('intrinsic' altruists) around.

Recent literature related to Chapter 10 includes Kane et al. (2014), Iacona et al. (2016), Lin and Li (2016), Sheng et al. (2017), Cumming (2018), Glynatsi et al. (2018) and Liu et al. (2018).

11 · *Incentive Design*

11.1 The Principal–Agent Problem

The present chapter takes up the problem of biodiversity loss as a public good problem (Section 9.1) and the possible solution of offering financial incentives for conservation, such as conservation payments. A conservation payment basically is a contract between a regulator and a landowner in which the landowner commits to carrying out conservation measures and the regulator pays a payment to the landowner that covers the associated costs. Both actors have interests: the regulator wishes for biodiversity to be conserved effectively and cost-effectively, and the individual landowner wants to maximise their profit.

The main challenge in the design of conservation payment schemes is that these interests often diverge and that information is usually distributed asymmetrically, for example in that the agent knows their own costs of conservation better than the conservation agency. To account for heterogeneity in the (unknown) conservation costs, the conservation agency can offer two (or more) contracts with different payment levels: higher payments for the high-cost landowners and lower payments for the low-cost landowners. The difficulty is that not all landowners will choose the appropriate contract, because the low-cost landowners in particular have an incentive to overstate their costs and pretend to be a high-cost landowner to collect the higher payment – a problem known as 'adverse selection' (Moxey et al. 1999; Laffont and Martimort 2002).

Another problem arises if the regulator cannot easily monitor compliance by the landowner with the contract. In this case the (self-interested) landowner has an incentive to omit the conservation measure but still pocket the payment. This is known as 'moral hazard' (Ozanne et al. 2001; Laffont and Martimort 2002) and, similarly to adverse selection, reduces the efficiency of the payment scheme.

Problems such as these can also occur in many other economic contexts such as employment and insurance contracts, and are termed principle–agent problems (Laffont and Martimort 2002). In a principal–agent problem a principal (the conservation agency, employer or insurance company) pays the agent (the landowner, employee, client) money for a certain action (conservation measure, labour) or in response to an insured incidence (e.g. car accident), respectively, and where a conflict exists between the interests of the principal and the agent.

At first sight the agent seems to be in a more powerful position than the principal, since the agent has more information than the principal and can freely decide on their actions while the principal is bound to the contract. However, the principal has an advantage that at least to some extent compensates for the informational disadvantage: that they can design the contract so that the agent, by acting in their own interest, also acts in the principal's interest. In the language of game theory, agent A (the principal) has the right to set the pay-offs of the various combinations of actions so that their own pay-off is maximised if agent B (the agent) chooses the action that maximises their own pay-off. In the context of conservation payments, the contract has to be designed so that a profit-maximising landowner automatically maximises the budget- or cost-effectiveness of the payment scheme. The challenge is to design such a contract and specify whom to pay how much for what, an economic research field termed 'mechanism design' (Diamantaras et al. 2009; Börgers et al. 2015). The following sections will outline a few examples of mechanism design for biodiversity conservation.

11.2 Payments under Asymmetric Information: Adverse Selection and Moral Hazard

In the context of conservation payments, the two classical problems of adverse selection and moral hazard have been analysed by Moxey et al. (1999) (with a variant by Smith and Shogren (2002)) and Ozanne et al. (2001), respectively. To model the problem of adverse selection I follow the simpler notation of Fees and Seliger (2013, ch. 11.4.2). Assume the conservation agency strives for a level of ecological benefit V that increases with the conservation effort executed by the landowner in a concave manner. In accordance with the definitions in Section 9.1 but without loss of generality, one may identify the conservation effort with the area A of conserved land. The benefit function is thus characterised by $\partial V/\partial A > 0$ and $\partial^2 V/\partial A^2 < 0$ (cf. Fig. 9.1, solid lines). The cost C of conservation increases with increasing conserved area A in a convex manner, that is, $\partial C/\partial A > 0$ and $\partial^2 C/\partial A^2 > 0$ (cf. Fig. 9.1, short-dashed lines).

To model heterogeneity among farms or land parcels, the magnitude of the marginal cost, $\partial C/\partial A$, is assumed to depend on some characteristic or type Θ of the landowner that is known to the landowner but not to the conservation agency. The type Θ may be the agricultural productivity of the land parcel which may be low, $\Theta = 1$, implying a low conservation cost (forgone profits), or high, $\Theta = h$, implying a high conservation cost.

If the agency knew the type Θ of each landowner, it could offer a homogenous payment p to all landowners so that each landowner would conserve an area A of an exact magnitude that leads to equality of payment and marginal cost, as prescribed by Eq. (9.13). This would maximise both the landowner's profit $p - C(A)$ and the agency's utility function

$$U(A) = V(A) - C(A) = V(A) - p. \qquad (11.1)$$

And, in particular, the agency would compensate the landowner exactly for their true cost. In addition, the allocation of payments would be cost-effective so that a low-cost landowner would conserve more land than a high-cost landowner (cf. Eq. 9.9).

The problem for the conservation agency is that it does *not* know which landowner is of type $\Theta = h$ and who is of type $\Theta = 1$. If in this case it offers a homogenous payment p, the low-cost landowners will pretend to be high-cost landowners so they have to conserve less land A for the offered payment than they would have to as low-cost landowners. The saved costs represent an income to the landowners, called an information rent, which represents a loss to the conservation agency.

How can the conservation agency solve this problem? A first-best solution in which the conservation agency has to pay zero information rent to the landowners does not exist, but the agency can minimise the information rents by offering two types of contracts: one with a payment p_l for conserving an area A_l, and one with a payment p_h for conserving an area A_h. The contracts have to be designed so that it is in the landowners' own interest to reveal their type, so that the high-cost landowners choose the contract with payment p_h and the low-cost ones choose the contract with payment p_l. Such a process is termed screening.

For the screening to be successful, four conditions have to be fulfilled:

$$p_l(A_l) \geq C_l(A_l), \qquad (11.2)$$

$$p_h(A_h) \geq C_h(A_h), \tag{11.3}$$

$$p_l(A_l) - C_l(A_l) \geq p_h(A_h) - C_l(A_h), \tag{11.4}$$

$$p_h(A_h) - C_h(A_h) \geq p_l(A_l) - C_h(A_l). \tag{11.5}$$

The first two inequalities, Eqs (11.2) and (11.3), express that for each landowner the payment has to exceed the cost. If these so-called participation constraints were not fulfilled the landowners would not participate in the payment scheme at all. The third inequality, Eq. (11.4), tells us that for the low-cost landowner it is more profitable to choose the low-cost contract (i.e. conserve A_l and receive a payment p_l) than the high-cost contract with A_h and p_h; and the fourth inequality, Eq. (11.5), states the analogous for the high-cost landowner. Consequently, if these two conditions are fulfilled, each landowner takes the contract that is intended for them. These conditions are termed incentive-compatibility constraints, to express that they ensure that profit-maximising agents behave just as desired by the principal.

To understand the functioning of this screening process one has to consider which of the four inequalities are binding, that is, in which the right-hand and left-hand sides are equal. For this assume that Eq. (11.3) is just fulfilled, so the payment $p_h(A_h)$ in the high-cost contract is exactly equal to the cost $C_h(A_h)$ of the high-cost landowner. This means that a high-cost landowner conserves exactly an area A_h at cost $C_h(A_h)$ which maximises their profit while avoiding the payment of any information rent by the agency.

However, as previously introduced, the conservation agency cannot a priori discriminate between high-cost and low-cost landowners and, in particular, cannot prevent a low-cost landowner from opting for the high-cost contract. A low-cost landowner choosing that option would behave like a high-cost landowner and conserve an area A_h to receive a payment p_h. But since the cost for the low-cost landowner is by definition lower than that for a high-cost landowner, the fulfilment of Eq. (11.3), $p_h = C_h(A_h)$, implies $p_h > C_l(A_h)$, that is, for the low-cost landowner the payment p_h in the high-cost contract strictly exceeds their associated costs $C_l(A_h)$. This means that the low-cost landowner is in the comfortable position of choosing between the low-cost and the high-cost contracts and, being a profit-maximiser, will opt for the high-cost contract to gain the information rent of magnitude $p_h - C_l(A_h)$.

The same can be said about the incentive-compatibility constraints. A high-cost landowner will never choose the low-cost contract, since it is associated with a lower payment p_l, so Eq. (11.5) is not binding, that is, it is automatically fulfilled. In contrast, a low-cost landowner benefits from mimicking a high-cost landowner to pocket the high payment, and Eq. (11.4) is binding.

The consequences of the information asymmetry can be seen from the agency's utility function, which is the sum of the net utilities, Eq. (11.1), obtained for all landowners. Assuming the agency knows the proportions q_l and q_h of low-cost and high-cost landowners, the (scaled) utility function reads

$$U = q_l[V(A_l) - p_l] + q_h[V(A_h) - p_h]. \tag{11.6}$$

Inserting Eqs (11.3) and (11.4) with equality of the respective left-hand and right-hand sides delivers

$$U = q_l[V(A_l) - C_l(A_l)] + q_h[V(A_h) - C_h(A_h)] - q_l[C_h(A_h) - C_l(A_h)]. \tag{11.7}$$

The first two terms represent the expected utility as the weighted differences between ecological benefits and conservation costs for the low-cost and high-cost landowners, respectively. They would also apply in the case of a perfectly informed conservation agency. The third term is due to the information asymmetry and represents the utility loss caused by the information rent that the conservation agency has to pay to the low-cost landowners.

The associated optimal payments p_l and p_h are determined by setting the derivatives of Eq. (11.7) with respect to A_l and A_h to zero, which yields

$$\frac{\partial V}{\partial A_l} = \frac{\partial C_l}{\partial A_l}, \tag{11.8}$$

$$q_h\left(\frac{\partial V}{\partial A_h} - \frac{\partial C_h}{\partial A_h}\right) = q_l\left(\frac{\partial C_h}{\partial A_h} - \frac{\partial C_l}{\partial A_h}\right). \tag{11.9}$$

For the low-cost landowner the optimality condition is the well-known equality between marginal benefit and marginal cost, which maximises the difference between benefit and cost, $V(A_l) - C_l(A_l)$. For the high-cost landowner, however, $\partial V/\partial A_h$ must be larger than $\partial C_h/\partial A_h$, because the right-hand side of Eq. (11.9) is positive by the definition of the cost functions. Since the benefit function is assumed to be concave and the cost functions are assumed to be convex, the positive difference $\partial V/\partial A_h - \partial C_h/\partial A_h > 0$ implies that the induced level of A_h is smaller than the level that would maximise the difference between benefit and cost, $V(A_h) - C_h(A_h)$.

The reason for this deviation is that the conservation agency is confronted with a trade-off between maximising the difference between benefit and cost, $V(A_h) - C_h(A_h)$, and minimising the information rent, $q_l[C_h(A_h) - C_l(A_l)]$: increasing the former will increase the latter. Equation (11.9) defines the optimal level of A_h where the sum of the efficiency loss associated with the reduced A_h and the information rent is minimised.

Now turn to the second major problem of incentive design: moral hazard. In the context of environmental problems it has been analysed, among others, by Ozanne et al. (2001). Translating the authors' main points into a conservation context, assume a landowner manages their land in a profit-maximising manner with associated profit π_0. Conserving an area of size A reduces the profit to a level of $\pi(A)$ where π is assumed to be decreasing with and concave in A: $\partial \pi/\partial A < 0$ and $\partial^2 \pi/\partial A^2 < 0$. By this, the cost $\pi_0 - \pi(A)$ incurred by conserving area A increases with and is convex in A, as in Fig. 9.1 (short-dashed lines). If the conservation agency wishes the landowner to conserve an area of some size A_+, it has to compensate for the forgone profit $\pi_0 - \pi(A_+)$ by a payment p of equal magnitude.

In the presence of moral hazard, however, the conservation agency does not know with certainty whether the landowner will really comply with the contract or not. However, one may assume that it can monitor the landowners' actions at a cost $M(q)$ where q is the probability of detecting non-compliance, and higher monitoring cost $M(q)$ implies a higher detection probability q. If the landowner is found guilty of non-compliance, they have to pay a fine of magnitude F.

Confronted with a payment of magnitude p for conserving an area A_+, the landowner has three choices:

1. do not participate in the scheme, and earn a profit of $\Pi_1 = \pi_0$;
2. participate and comply with the scheme, and earn a profit of $\Pi_2 = p + \pi(A_+)$;
3. participate in the scheme and not comply, and earn a profit of $\pi_0 + p$ in the case of non-detection and $\pi_0 - F$ in the case of detection, so that the expected profit equals $\Pi_3 = (1 - q)(\pi_0 + p) + q(\pi_0 - F)$.

To induce participation in the scheme by a profit-maximising landowner, the participation constraint $\Pi_2 \geq \Pi_1$ must be fulfilled, and to secure landowner compliance in the scheme requires fulfilment of the incentive-compatibility constraint $\Pi_2 \geq \Pi_3$.

Given these two constraints and a given level for the fine F (which is set by law and cannot be chosen by the agency), the conservation agency has the task to select the level of conserved area A_+, the payment p and the monitoring effort (detection probability) q to maximise its utility

$$U(A_+, p, q) = V(A_+) - p - M(q) \qquad (11.10)$$

which is the difference between the conservation benefit $V(A_+)$ associated with conserved area A_+, and the agency's expenses to run the scheme, p and $M(q)$. Assuming that V is increasing with and concave in A so that $\partial V/\partial A > 0$ and $\partial^2 V/\partial A^2 < 0$ (cf. Fig. 9.1, solid lines), the scheme parameters that maximise U are given by (cf. Ozanne et al. 2001)

$$\left\{ 1 + \frac{\partial M/\partial q(\pi_0 + p*)F}{p* + F} \right\} \cdot \left. \frac{\partial \pi}{\partial A_+} \right|_{A_+ = A_+^*} + \left. \frac{\partial V}{\partial A_+} \right|_{A_+ = A_+^*} = 0, \qquad (11.11)$$

$$p* = \pi_0 - \pi(A_+), \qquad (11.12)$$

$$q* = \frac{p*}{p* + F}. \qquad (11.13)$$

According to Eq. (11.11), if monitoring could be carried out at zero cost, $M = \partial M/\partial q = 0$, the optimal level of conservation, A_+^*, would be given by the well-known equality of marginal conservation benefit and marginal conservation cost: $\partial V/\partial A|_{A_+*} = -\partial \pi/\partial A|_{A_+*}$. Considering non-zero costs of monitoring ($M > 0$ and $\partial M/\partial q > 0$), in contrast, increases the content in the braces in Eq. (11.11), which implies (since $\partial V/\partial A|_{A_+*}$ is fixed) that $|\partial \pi/\partial A|_{A_+*}$ must decrease. Since π is decreasing and concave in A ($\partial \pi/\partial A < 0$ and $\partial^2 \pi/\partial A^2 < 0$), a smaller level $|\partial \pi/\partial A|_{A_+*}$ implies a smaller conserved area A_+^* compared with the first-best situation in which monitoring incurs no cost.

Since A_+^* is reduced, the optimal payment $p* = \pi_0 - \pi(A_+^*)$ (Eq. 11.12) is reduced compared with the first-best situation as well. Lastly, the optimal level of monitoring effort $q*$ (Eq. 11.13) increases with decreasing fine F, because a higher fine deters landowners from non-compliance. It further increases with increasing payment $p*$ and thus declines with increasing monitoring cost.

To summarise, large (marginal) monitoring costs imply a lower optimal conservation effort, a lower associated payment, lower optimal monitoring effort and a lower detection probability of non-compliance.

11.3 Cost-Effectiveness versus Budget-Effectiveness

As previously discussed, for the conservation agency it would be ideal if it could compensate each landowner i exactly according to their actual conservation cost c_i. Otherwise, under homogenous payments where each landowner receives the same payment p, landowners with costs $c_i < p$ pocket an information rent of magnitude $p - c_i$. Within the framework of demand and supply (cf. Fig. 9.1b), this information rent is also termed the landowner's producer rent, which is the difference between the revenue p of selling a good (conserved area) on the market and the production cost c_i. Summed over all landowners, the producer rent is represented by the light grey area between the horizontal payment line and the marginal cost curve in Fig. 11.1.

The total production cost is the integral of the marginal cost (cf. Eq. 9.4), that is, the area under the marginal cost curve (Fig. 11.1, dark grey). The total conservation expense, or conservation budget, is the rectangular area under the horizontal payment line, pA. This means that the conservation budget is the sum of the landowners' total conservation costs and their producer rents.

A conservation instrument is cost-effective if the total conservation cost is minimised, which is achieved if the landowners with lower costs conserve more than landowners with higher costs (Eq. 9.9), or – assuming each landowner has only the choice between conserving an area a or not as in Fig. 11.1 – if the landowners with the low costs $c_i < p$ conserve

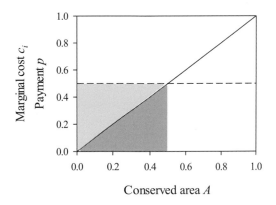

Figure 11.1 Regional marginal cost c_i, scaled between zero and one (solid line), and payment p (dashed line) as functions of conserved area A, scaled between zero and one, in the model region (cf. Fig. 9.3). Light and dark grey colours represent the producer rent and the conservation cost, respectively, for a conserved area of size 0.5.

and the others use their land for economic purposes. But what is the effect of the producer rent here? From the point of view of social welfare, which measures the difference between the sum of all social benefits and the sum of all social costs, the magnitude of the producer rent is irrelevant, since it only represents a transfer of money from one part of the society – the taxpayers who eventually have to finance the conservation payments – to the landowners.

This is, however, only partly true. Raising taxes to finance public goals incurs transaction costs, since taxes have to be collected, funds have to be managed and so forth. Secondly, any fiscal transfer between economic sectors bears the risk that money is shifted from a more productive to a less productive sector of the economy, implying a loss of social welfare. This is known as tax distortion (Laffont and Tirole 1993, p. 55; Innes 2000). Although producer rents do have economic advantages, since, for example, they reward innovation towards higher productivity (also with regard to generating conservation benefits), it is generally desirable to keep producer rents low and minimise the conservation budget, that is, the sum of conservation costs and producer rents. In this way, *budget-effectiveness* is said to be maximised.

It is important to be clear about whether the aim of incentive design is to maximise cost-effectiveness or budget-effectiveness, because different conservation instruments may perform differently with regard to these two criteria. Drechsler (2017b) compared the performances of the agglomeration payment of Drechsler et al. (2010) outlined in Section 13.3 with the spatially targeted auction mechanism of Polasky et al. (2014) outlined in Section 13.5 with regard to their cost-effectiveness and budget-effectiveness and found that the agglomeration payment outperforms the auction with regard to budget-effectiveness but is outperformed with regard to cost-effectiveness.

11.4 Auctions as an Alternative to Address Adverse Selection

An alternative and increasingly popular instrument that forces agents or landowners to reveal information about their true conservation costs are reverse auctions. The term 'reverse' is used to indicate that in contrast to an ordinary auction, the bidders (landowners) express not the payment they are willing to pay for an object but the payment they demand from the buyer (the conservation agency) of the object. In the context of conservation, in which multiple agents offer a good (conserved land), discriminatory first-price sealed-bid auctions are the most common and appropriate (Latacz-Lohmann and van der Hamsvoort 1997; Reeson et al. 2011). The term 'sealed bid' means that the bidders only know their own bids and not those of the other landowners; 'first-price' means that the landowner receives exactly the payment demanded for the conserved land; and 'discriminatory' means that each landowner receives their own individual bid – in contrast to uniform-price auctions in which all bidders receive the same payment.

Since in such an auction only the lowest bids (demanded payments) are accepted by the agency, each landowner has an incentive not to overstate their costs and will avoid submitting an excessively high bid. As a profit-maximiser, similar to a blackjack player, they are confronted with the trade-off between demanding as high a payment as possible and minimising the risk of overbidding. To demonstrate the consequences of this trade-off for the functioning of a conservation auction, I follow Latacz-Lohmann and van der Hamsvoort (1997) and assume a number of N landowners i ($i = 1, \ldots, N$) with differing conservation costs c_i. Setting, without loss of generality, the profit of economically used land to zero, if a landowner submits a bid b_i that is accepted by the conservation agency the landowner gains a profit of

$$\pi_i = b_i - c_i. \tag{11.14}$$

This is obviously positive only if the bid exceeds the conservation cost, so a landowner will never bid less than their costs (participation constraint). In order to keep to its budget constraint, the conservation agency sets a maximum acceptable bid β, so that bids above β are rejected. If that bid cap were known to the landowners, all landowners would of course demand β, but since β is private information known only to the conservation agency, they can only guess at the level of β. If $f(\beta)$ is denoted as the subjective probability density of β (i.e. the landowners' beliefs about the level of β), the probability of a bid b_i being below β is

$$\Pr(b_i < \beta) = \int_{b_i}^{\infty} f(\beta)d\beta. \tag{11.15}$$

The expected profit associated with a bid of magnitude b_i thus is

$$\Pi_i = \Pr(b_i < \beta) \cdot (b_i - c_i). \tag{11.16}$$

The profit-maximising bid b_i^* is found by setting the derivative of Π_i with respect to b_i to zero, which (with $\partial \Pr(b_i < \beta)/\partial b_i = -f(\beta = b_i)$) yields

$$b_i* = c_i + \frac{\Pr(b_i* < \beta)}{f(\beta = b_i*)}. \tag{11.17}$$

For the special case of β being uniformly distributed in the interval $[\beta_{min}, \beta_{max}]$, Eq. (11.17) can be solved for b_i^* (Latacz-Lohmann and van der Hamsvoort 1997):

$$b_i* = \max\{(c_i + \beta_{max})/2, \ \beta_{min}\} \quad \text{s. t.} \quad b_i* \geq c_i. \tag{11.18}$$

Equation (11.18) implies that the profit-maximising bid increases with increasing conservation cost c_i and increasing expected bid cap.

The interesting question now is, how much higher is the budget-effectiveness of this auction scheme compared with a homogenous payment? To simplify the analysis I assume a large (mathematically precisely: infinite) number of landowners whose conservation costs are distributed uniformly in the interval $[1 - \sigma, 1 + \sigma]$. The assumption of an infinite number of landowners allows for treating the conservation cost c as a continuous variable where

$$q(c) = \begin{cases} (c - 1 + \sigma)/(2\sigma) & 1 - \sigma \leq c \leq 1 + \sigma \\ 0 & \text{otherwise} \end{cases} \tag{11.19}$$

is the proportion of landowners with cost below c; and equivalently, if q is the proportion of the landowners with the lowest costs, then

$$c(q) = 1 - \sigma + 2q\sigma \tag{11.20}$$

is the highest cost observed among that set of low-cost landowners (Drechsler 2011). In other words, one can regard the costs of all land-owners arranged in increasing order from $1 - \sigma$ to $1 + \sigma$ so that each landowner is characterised by a particular value q (and a proportion q of landowners having lower costs and $1 - q$ have higher costs). By varying q from zero to one, all landowners from the lowest-cost type $(1 - \sigma)$ to the highest-cost type $(1 + \sigma)$ are addressed.

Under a homogenous payment p, the proportion q of landowners who have conservation cost c below p and conserve their land thus is

$$q(c) = \begin{cases} 0 & p \leq 1 - \sigma \\ (p - 1 + \sigma)/(2\sigma) & 1 - \sigma \leq p \leq 1 + \sigma \\ 1 & p \geq 1 + \sigma \end{cases} \tag{11.21}$$

and the payment required to induce a proportion q of landowners to conservation is

$$p(q) = 1 - \sigma + 2q\sigma. \qquad (11.22)$$

The *budget* required to finance the payment that goes to the proportion q of conserving landowners is

$$B(q) = Nqp(q) = N[(1 - \sigma)q + 2\sigma q^2]. \qquad (11.23)$$

For comparison, the *total conservation cost* of the q landowners is (cf. Eq. 9.4)

$$C_{tot}(q) = N \int_{1-\sigma}^{1+\sigma} p(q')dq' = N\left[(1 - \sigma)q + \sigma q^2\right] \qquad (11.24)$$

While Eq. (11.23) represents the budget for homogenous payments, under heterogeneous payments p_i ($i = 1, \ldots, N$) where each landowner is paid exactly according to their cost, $p_i = c_i$, the agency's budget equals exactly the total cost C_{tot} of Eq. (11.24).

The difference between B of Eq. (11.23) and C_{tot} of Eq. (11.24) equals σq^2, implying that the budget under homogenous payments exceeds that under heterogeneous payments by an amount σq^2 – which represents the landowners' producer rent discussed in Section 11.3. Now turn to the auction scheme. In accordance with Latacz-Lohmann and van der Hamsvoort (1997), I assume that the landowners believe the bid cap β to lie within an interval $[1 - \gamma, 1 + \gamma]$ around the mean conservation cost which had been set to 1. Since it is not sensible to expect a bid cap above the largest possible conservation cost or below the lowest cost, we have $\gamma \leq \sigma$. With these assumptions, the profit-maximising bid $b^*(c)$ becomes

$$b^*(c) = \max\left\{\frac{(c + 1 + \gamma)}{2}, \; 1 - \gamma\right\} \qquad (11.25)$$

with the participation constraint that the bid must exceed the conservation cost: $b^* \geq c$. Inserting Eq. (11.20), the profit-maximising bid of landowner q becomes

$$b^*(q) = \max\left\{\frac{1 + (\gamma - \sigma)}{2} + q\sigma, \; 1 - \gamma\right\} \quad \text{s.t.} \quad b^*(q) \geq 1 - \sigma + 2q\sigma.$$

$$(11.26)$$

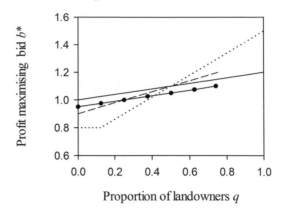

Figure 11.2 Profit-maximising bid b^* as a function of landowner index q. Lines without circles: $\gamma = 0.2$ with $\sigma = 0.2$ (solid line), $\sigma = 0.4$ (dashed line) and $\sigma = 0.8$ (dotted line); line with circles: $\gamma = 0.1$, $\sigma = 0.2$. For the dashed line and the line with circles the participation constraint, Eq. (11.27), is binding.

Two cases have to be distinguished. For $\sigma \leq 3\gamma$ the profit-maximising bid is given by the first argument of the max function in Eq. (11.26) with the participation constraint

$$q \leq q_c \equiv \frac{\gamma + \sigma}{2\sigma}. \qquad (11.27)$$

This means that $b^*(q)$ increases with increasing q (conservation cost) up to a certain maximum (given by Eq. (11.27)) beyond which $b^*(q)$ does not lead to a positive expected profit (Fig. 11.2, solid and dashed lines and line with circles). The slope of that increase, $\partial b^*/\partial q$, is given by the level of cost variation σ while an increase in γ lifts b^* up by an amount $\gamma/2$ for all q.

For the other case, $\sigma > 3\gamma$, there exists a threshold value

$$\hat{q} = \frac{\sigma - 3\gamma}{2\sigma} > 0, \qquad (11.28)$$

such that for $q \geq \hat{q}$ we obtain the same result as for the above case of $\sigma \leq 3\gamma$, while for $q \leq \hat{q}$ the profit-maximising bid is constant at $b^* = 1 - \gamma$ (observed for $q \leq 0.125$ in the dotted line of Fig. 11.2), with the participation constraint not being binding. The reason for the constancy of b^* for small q is that small q imply that only low-cost landowners bid – and all maximise their expected profit by bidding at the smallest believed bid cap of $1 - \gamma$.

The fact that all landowners submit the same bid is problematic for the conservation agency, because it is impossible to discriminate between the proportion q of bids it wants to accept and the proportion $\hat{q} - q$ it wants to reject. The agency can only decide on an ad hoc basis – which would violate the initial premise that *all* bids below the bid cap are accepted. The reason for this problem is the wrong expectations of the landowners regarding the level of the bid cap: a bid cap near the mean conservation cost of 1 (as is implied by the comparatively small γ) is simply not consistent with a small proportion q of accepted bids, since small q are associated with small conservation costs far below 1.

To assess the budget-effectiveness of the auction compared with that of homogenous (Eq. 11.23) and differentiated payments (Eq. 11.24) I calculate the budget $B^*(q)$ through integration of Eq. (11.26),

$$B^*(q) = N \int_0^q b^*(q')dq' \quad \text{s.t.} \quad q \le q_c, \tag{11.29}$$

and plot $B(q)$, $C(q)$ and $B^*(q)$ as functions of q. Figure 11.3 shows that the budget-effectiveness is, naturally, highest under differentiated payments for all q, while the auction outperforms the homogenous payment for q larger than about 0.5 and underperforms for smaller q.

It may seem surprising initially that the homogenous payment can outperform the auction at all. However, this observation can easily be explained by the fact that at very small q the associated conservation costs are close to $1 - \sigma$ and so is the homogenous payment that equals the maximum of those relatively small costs. As a consequence, the required budget is roughly $q(1-\sigma)$. In the auction scheme, the smallest possible bid is $1 - \gamma$, so for very small q the budget is roughly $q(1 - \gamma)$, which is larger than the budget for the homogenous payment. The reverse is found at large $q \approx 1$. Here the homogenous payment is close to the maximum cost, $1 + \sigma$, leading to a budget of around $1 \cdot (1 + \sigma)$. The average profit-maximising bid for $q \approx 1$ is somewhat below 1 (Fig. 11.3) and so is the required budget, which therefore is below the budget for the homogenous payment.

What is the principle reason for the auction underperforming at low q? Similar to the previous discussion, the reason is a wrong expectation of the bid cap β. For small q and implied conservation costs around $1 - \sigma$, a bid cap larger than $1 - \gamma$ (with $\gamma < \sigma$) is simply not plausible; instead a reasonable guess on β here would be close to the relevant conservation costs, that is, a little above $1 - \sigma$. One can conclude that the present auction model leads to sensible results only if the expected bid cap is reasonable, which in the present parameterisation is the case only for medium or large q.

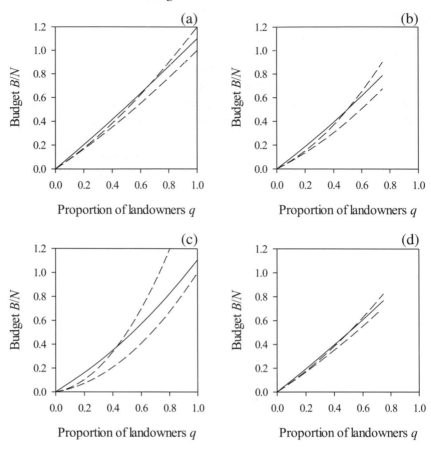

Figure 11.3 Scaled budget B/N associated with a proportion q of landowners conserving their land under homogenous payments (upper dashed line), differentiated payments (lower dashed line) and the auction scheme (solid line) for the parameter combinations of Fig. 11.2: $\gamma = 0.2$ $\sigma = 0.2$ (panel a), $\gamma = 0.2$, $\sigma = 0.4$, (panel b), $\gamma = 0.2$, $\sigma = 0.8$ (panel c), and $\gamma = 0.1$, $\sigma = 0.2$ (panel d).

11.5 Payments for Heterogeneity

Another example where payments are used to induce differentiated behaviour in landowners can be found in Ohl et al. (2008). This paper is motivated by the observation that biodiversity in cultural landscapes often hinges on the diversity of land use. Many birds and butterflies in Central Europe rely on grasslands. On the one hand, these grasslands need to be kept open by mowing or grazing to provide suitable conditions for reproduction and feeding. On the other hand, many of the grassland species can survive only if the land use, for example mowing, is carried out outside the reproductive period. These reproductive periods differ greatly among birds and butterflies, and even among different butterfly species, so that within the whole summer season no single point in time exists at which a cut does not harm at least one species. This implies that biodiversity can be maintained only if the grassland is managed as a mosaic in which different patches of grassland are mowed at different times.

This circumstance was the motivation to develop the software DSS-Ecopay mentioned in Section 2.7. In that software, land-use diversity can be induced through a set of payments – one for each land-use measure, because the costs of the different land-use measures differ among grassland parcels. If each landowner chooses the land-use measure that maximises the difference between payment and cost, a mosaic of grassland parcels with different land-use measures is induced. Whether a desired land-use mosaic can be induced through a set of payments, however, is not trivial and depends heavily on the cost structure in the landscape.

To demonstrate this, Ohl et al. (2008) considered the simple case of three land parcels, each managed by one landowner. The conservation agency's objective is to induce two of the landowners to carry out conservation measures – some measure A by one landowner and measure B by the other – while the third landowner should use their land parcel conventionally. To achieve that objective the agency can set two payments, p_A and p_B, that are offered to the landowner who carries out measure A and measure B, respectively.

Since the intention is for one of the landowners to use their land conventionally, it does not make much sense to distinguish between participation and incentive-compatibility constraints as above, but just speak of incentive-compatibility constraints. These are obtained by demanding that for each landowner, exactly one measure maximises the landowner's profit, that is, the difference between payment and conservation cost. Denoting the conservation cost of measure m ($m \in$ {A,B}) on land parcel i as $C_i(m)$, the three incentive-compatibility constraints can be written as (Ohl et al. 2008):

(a) $C_2(A) < p_A < C_1(A)$
(b) $C_3(B) < p_B < C_1(B)$. (11.30)
(c) $C_3(B) - C_3(A) < p_B - p_A < C_2(B) - C_2(A)$

This set of inequalities can be solved graphically to obtain a necessary and sufficient condition for the existence of a payment scheme $\{p_A, p_B\}$ that is able to induce the desired land–use mosaic. As Ohl et al. (2008) found, to test whether Eq. (11.30) has a solution, the sum $C_i(A) + C_j(B)$ must be formed for all six landowner combinations ($i, j = 1, \ldots, 3; i \neq j$), and only if one of the six sums is strictly smaller than the other five sums, that is, if and only if some $u \neq v$ exist such that

$$C_u(A) + C_v(B) < \min_{\substack{i,j \in \{1,2,3\} \\ i \neq j \\ (i,j) \neq (u,v)}} \{C_i(A) + C_j(B)\}, \qquad (11.31)$$

then a payment scheme exists that solves Eq. (11.30) and induces complete land-use heterogeneity. Otherwise, at least two landowners carry out the same land-use measure.

The case of three land parcels can be extended to cases of $N > 3$ land parcels by demanding that land parcels i with $i \in I_A = \{1, \ldots, N_A\}$ are managed with measure A, land parcels $i \in I_B = \{N_A+1, \ldots, N_A+N_B\}$ are assigned measure B and the remaining land parcels $i \in I_0 = \{N_A+N_B+1, \ldots, N\}$ are managed conventionally. This land-use pattern can be induced by payments p_A and p_B if and only if

$$C_{tot} = \sum_{i \in I_A} C_i(A) + \sum_{i \in I_B} C_i(B) < \min_{\substack{(I_A' \neq I_A) \vee (I_B' \neq I_B), \\ |I_B'| = |I_B|, \, |I_B'| = |I_B|}}$$

$$\left\{ \sum_{i \in I_A'} C_i(A) + \sum_{i \in I_B'} C_i(B) \right\}. \qquad (11.32)$$

To identify through Eq. (11.32) whether heterogeneous land use can be induced, all possible allocations of land parcels must be considered where exactly $N_A \equiv |I_A|$ land parcels are managed with measure A and $N_B \equiv |I_B|$ land parcels are managed with measure B, and there must be a single allocation that is associated with strictly lower total cost C_{tot} (sum over the conservation costs of all land parcels) than all other allocations. If this condition is not fulfilled, measure A is applied on fewer or more land parcels than N_A and/or measure B is applied on fewer or more land parcels than N_B. Analyses by Ohl et al. (2006) suggest that Eq. (11.32) is fulfilled if the rank order of costs $C_i(A)$ ($i = 1, \ldots, N$) equals the rank order of costs $C_i(B)$, such that if the cost $C_i(A)$ of measure A on land parcel i is the y-th highest among all land parcels, the cost $C_i(B)$ of measure B is the y-th highest, as well, for all $y \in N$. Or in other words, the cost functions $C_i(m)$ of different land parcels as functions of m must not cross each other.

11.6 The Agglomeration Bonus

In the previous examples of payment design the spatial location of the conservation measures or the conserved land played no role. However, in Chapter 6 it was shown that the viability of a spatially structured population increases with increasing ability of the individuals to disperse between habitat patches, which is usually negatively related to the distances between the habitat patches. In the design of biodiversity conservation instruments it would therefore be ecologically beneficial if these could ensure that conservation measures and species habitat are spatially agglomerated rather than dispersed.

A design concept to achieve this has been proposed by Smith and Shogren (2001) and Parkhurst et al. (2002), who suggest supplementing the homogenous conservation payment (which they term 'base payment') with a bonus payment if the conserved habitat borders on another habitat – whether it is owned by the same individual or another landowner. The more joint borders the habitat has with other habitats (in the case of a square grid as considered by the authors: the more adjacent grid cells are conserved), the higher the bonus.

Analogous to the authors' arguments, to demonstrate the effect of such a bonus on the land-use pattern consider four land parcels arranged in a row where the two western parcels belong to some landowner A and the two eastern ones to some landowner B. The costs of the land parcels are given in Table 11.1.

Assume a conservation agency offers a payment of magnitude €300 for a conserved land parcel, and for didactic purposes (for a more realistic setting, see Parkhurst et al. (2002)) assume that each landowner can conserve at maximum one land parcel. If the two landowners are rational profit-maximisers they will conserve the two outer land parcels A1 and B2. Now let the conservation agency offer an additional bonus of €200 for a conserved land parcel that adjoins another conserved land parcel. Since for the given base payment each landowner conserves exactly one land parcel and the question is only whether this is an inner or an outer

Table 11.1 *Conservation costs of four land parcels arranged in a row.*

Parcel A1	Parcel A2	Parcel B1	Parcel B2
€250	€400	€400	€250

Table 11.2 *Pay-off of landowner A dependent on the pattern of conserved land parcels.*

		Landowner B	
		Parcel B1 conserved	Parcel B2 conserved
Landowner A	Parcel A2 conserved	$300 - 400 + 200 = 100$	$300 - 400 = -100$
	Parcel A1 conserved	$300 - 250 = 50$	$300 - 250 = 50$

one, the landowners have altogether four choices, whose pay-offs are given in Table 11.2.

Comparison with Table 10.2 reveals that this pay-off structure represents a coordination game with two Nash equilibria: one in which the two outer land parcels, A1 and B2, are conserved, and one in which the two inner ones, A2 and B1, are conserved. The latter equilibrium is the dominant one which leads to the higher pay-off for both landowners. Thus, under the considered agglomeration bonus scheme, two rational landowners will conserve the two connected inner land parcels although they have higher conservation costs.

The conceptual economic explanation of this finding is that under the agglomeration bonus conservation creates a positive spatial externality for neighbouring landowners that makes conservation of adjacent land parcels more rewarding for them. This is similar to the case described in Table 2.1 of Section 2.4, where two countries can choose between conserving a forest and developing the land. In that example the forest conserved by country B provides a positive externality to country A, and that spatial externality from the neighbour's forest can be enjoyed even if the own forest is not conserved, leading to the described prisoner's dilemma with the single Nash equilibrium. In contrast, in the agglomeration bonus scheme the spatial externality can be enjoyed only if the own inner land parcel is conserved. Therefore if the neighbour cooperates and conserves their inner land parcel, 'defection' by conserving the less costly outer land parcel is disadvantageous – turning the prisoner's dilemma into a coordination game with the two mentioned Nash equilibria.

The critical practical question is whether players are able to find the dominant Nash equilibrium. For the coordination problem of Table 11.2 there is no doubt about that, but in the more realistic setting by Parkhurst et al. (2002) with altogether 15 land parcels per player, each of which can be conserved or not, the answer is highly non-trivial. However, in an experiment the authors found that in the vast majority of trials the players were able to find the dominant (agglomeration) Nash equilibrium, especially if the players were allowed to communicate with each other. The economics of the agglomeration bonus will be presented in more detail in Chapter 13.

Recent literature related to Chapter 11 includes Derissen and Quaas (2013), Lennox et al. (2013), de Fries and Hanley (2016), van Kooten and Johnston (2016), White and Hanley (2016), Choi et al. (2018) and Sheng and Qiu (2018).

12 · *Modelling Human Decisions*

In the modelling of human decisions, three questions need to be addressed: what are the alternative strategies and actions from which a most favourable one is to be selected, what are the objectives and preferences along which the alternatives are evaluated, and against which informational background does this evaluation take place. The available set of alternatives is rather problem-specific and not related to the modelling of human behaviour, so it is not considered in the present chapter. Objectives will be dealt with in three sections, focusing on three relevant aspects: the concept of trade-offs and diminishing returns, decisions under uncertainty and equity preferences. Next, three sections will address the issues of imperfect information and bounded rationality. A final section will deal with the modelling of the change of actions and strategies in a dynamic setting.

12.1 Utility Functions

In the previous three chapters, agents were largely modelled in a rather simplistic manner: as self-interested rational profit-maximising individuals that are perfectly informed about the consequences of their decisions. The following sections will depart from this simple model and gradually add more complexity and realism.

In Sections 9.1 and 11.2 we used utility functions to model the favourability of a policy for society or for a conservation agency as society's representative. In an analogous manner, utility functions can also be used to model decisions of individual agents. A utility function basically provides a preference order of actions and their outcomes that are ranked with regard to their relative favourability to the decision-maker. Considering the outcome 'profit' π as the difference between benefits and costs, the simplest utility function is the linear function $U(\pi) = \alpha\pi$ with some positive constant α. Assuming that rational individuals maximise their utility U, they will thus choose the action that maximises their profit π.

This linear utility function assumes that each additional unit of profit contributes equally to utility, regardless of whether the current profit is already large or small. This contrasts with the common observation of decreasing marginal utility, which means that the contribution of an additional unit of some good (such as cars or chocolate) usually declines with increasing amount x of that good already owned. To model decreasing marginal utility, the utility function must be concave with $\partial U(x)/\partial x > 0$ and $\partial^2 U(x)/\partial x^2 < 0$, like the benefit function shown in Fig. 9.1. Typical parametric examples of concave utility functions – often chosen for mathematical tractability – are $U(x) \sim 1 - \exp(-\alpha x)$, $U(x) \sim x^{\alpha}$ with $\alpha < 1$, and $U(x) = \alpha x - \beta x^2$ for $x \leq \alpha/2\beta$.

Since these utility functions are strictly monotonous ($\partial U/\partial x > 0$), the maximisation of U and x lead to the same result for the chosen optimal decision alternative x. However, in cases where an individual has several objectives the shape of the utility function, in particular its concavity, can have a strong effect on the chosen action. To demonstrate this, consider an individual whose well-being is determined by the consumption of two goods: number x of textbooks on ecological modelling, and number y of textbooks on economic modelling. Assuming that each additional ecological-modelling book increases total utility by α and each economic-modelling book increases it by β, total utility can be written as

$$U(x, y) = \alpha x + \beta y. \qquad (12.1)$$

The dependence of U on x and y can be conveniently depicted in an iso-utility diagram (Fig. 12.1a). Each level of utility is represented by a straight line with slope $-\alpha/\beta$, and all points on the line are associated with the same utility. The negative slope indicates that a loss of ecological-modelling books by a number Δx can be compensated for by a gain in economic-modelling books by $\Delta y = \alpha/\beta \times \Delta x$. The iso-utility line is straight, which means that the substitution rate between the two goods, $\Delta y/\Delta x$, is constant and independent of the current endowment of ecological-modelling and economic-modelling books.

Alternatively, assume decreasing marginal utility in both goods, such that U is given by

$$U(x, y) = \alpha x^{0.5} + \beta y^{0.5}. \qquad (12.2)$$

The corresponding iso–utility lines are shown in Fig. 12.1b. In contrast to Fig. 12.1a, the substitution rate $\Delta y/\Delta x$ is not constant but depends on the

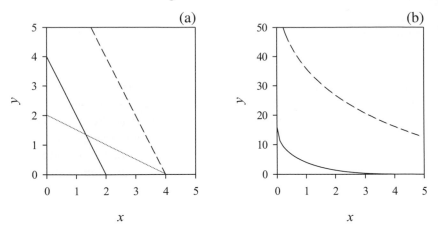

Figure 12.1 Iso-utility lines for the utility functions of Eqs (12.1) (panel a) and (12.2) (panel b). Parameters in panel a: $U = 4$, $\alpha = 2$, $\beta = 1$ (solid line), $U = 8$, $\alpha = 2$, $\beta = 1$ (dashed line), $U = 4$, $\alpha = 1$, $\beta = 2$ (dotted line); panel b: $\alpha = 2$, $\beta = 1$, and $U = 4$ (solid line) and $U = 8$ (dashed line) (note the different scales of the x- and y-axes).

current endowment of the two goods: the larger the current number of ecological-modelling books, x, compared with the number of economic-modelling books, y, the smaller the substitution rate $\Delta y/\Delta x$, which means that only a few additional economic-modelling books y can compensate for a large number of lost ecological-modelling books x; the converse can be said if y is large compared with x.

An alternative interpretation of the difference between the shapes of the iso-utility lines in Figs 12.1a and 12.1b is by the levels of substitutability and complementarity they represent. In Fig. 12.1a, ecological-modelling books can be perfectly substituted by economic-modelling books, that is, a loss in one good can be fully compensated for by an adequate gain in the other according to the substitution rate $\Delta y/\Delta x$. In Fig. 12.1b, in contrast, substitution is restricted so that at large x/y a large gain Δx is required to compensate for even a small loss $-\Delta y$. In the extreme, that is, at very ('infinitely') large x/y, a loss in y cannot be compensated for by gains in x at all. This means that the two types of books are (to some degree) complementary in that the presence of one good is required to draw any (marginal) utility from the other. In the present example, an additional ecological-modelling book increases total utility measurably only if there are sufficiently many economic-modelling books around. The spectrum between perfect substitutability and perfect complementarity can be conveniently modelled by the utility function

$$U(x, y) = (\alpha x^{1-\gamma} + \beta y^{1-\gamma})^{1/(1-\gamma)}, \qquad (12.3)$$

where α and β, as previously, represent the preferences or weights of the two goods x and y, and $\gamma = 0$ represents perfect substitutability while $\gamma \to \infty$ represents perfect complementarity (Quaas et al. 2013). The utility functions of Eqs (12.1) and (12.2) represent special cases of Eq. (12.3) with $\gamma = 0$ and $\gamma = 0.5$, respectively (note that in the latter case, Eq. (12.2) is actually missing a squaring down to obtain Eq. (12.3), but since the strictly monotonic mapping $x \to x^2$ does not change the relative rankings of the utilities $U(x,y)$, both functions – with and without squaring down – are economically equivalent). Of course, all of these demonstrations and discussions can be easily extended to more than two goods.

A final issue that is relevant in the context of utility functions is the common situation in which profits or utilities do not accrue at a single point in time but over several time steps, $t' = 0, \ldots, t$. The standard economic approach here is to aggregate these time-dependent utilities $u(t')$ to a single present-value utility by multiplying the temporal utilities with a discount factor $1/(1+r)^{t'}$ that decreases with increasing discount rate r and increasing time t', so that utilities accruing in the later future are valued less than utilities in the present or the near future. The strength of that decline is measured by the discount rate r. The resulting products of utility and discount factor are added to obtain the present-value utility. In discrete and continuous time they read, respectively:

$$U(t) = \sum_{t'=0}^{t} \frac{u(t')}{(1+r)^{t'}}, \qquad (12.4)$$

$$U(t) = \int_0^t u(t')e^{-rt'}dt'. \qquad (12.5)$$

Why are utilities accruing in the near future given higher weight than utilities from later time periods? Major reasons for this include impatience, economic growth which makes a euro less valuable next year than this year and the risk that future consumption may be inhibited, for example due to the death of the individual (Frederick et al. 2002). A more detailed discussion of the concept of discounting is far beyond the scope of the present book. Discounting is a critical step in the economic valuation of environmental goods such as biodiversity (cf. Sections 16.5–16.8).

12.2 Risk–Utility Functions: Modelling Decisions under Risk

The consequences of actions such as managing land for conservation are often uncertain. Economists distinguish between the notion of risk, under which all possible outcomes and their probabilities are known, and uncertainty *sensu stricto*, where the possible outcomes are known but their probabilities are unknown (Knight 1921). The following presentations will assume situations of risk.

The presence of risk and uncertainty can decisively affect our decisions. Most humans are risk-averse, which means, for example, that they prefer a certain income of €100 to a lottery in which they gain, say, €50 and €150 with probability 0.5 each. Although the *expected* pay-off is the same under both alternatives, the pay-off in the second situation is variable. To model aversion to this sort of variability one can, for example, build a utility function that includes not only the expected value μ but also the standard deviation σ of the pay-off:

$$U(\mu, \sigma) = \mu - k\sigma. \tag{12.6}$$

A value of $k = 0$ here models risk neutrality under which only the expected pay-off counts, so that the two alternatives above, €100 versus €50 or €150, are valued equally. A positive k implies that utility declines with increasing variability σ. Increasing $k > 0$ models increasing degree of risk aversion and the certain €100 becomes increasingly preferred to the lottery. Negative $k < 0$, in contrast, characterises risk-seeking behaviour which, for example, explains gambling behaviour. A convenient application of this utility function is that outcomes of different actions can be plotted in μ–σ space, and the iso-utility lines, $U(\mu,\sigma) =$ const., allow for identifying the utility-maximising actions graphically (Fig. 12.2; cf. Quaas et al. 2007). In particular, one can see that an increase in risk can be compensated for by an according increase in the expected pay-off.

More common in the modelling of human decisions are utility functions that relate utility, as described in Section 12.1, to some outcome such as profit. In the context of risk modelling, the concave shape of

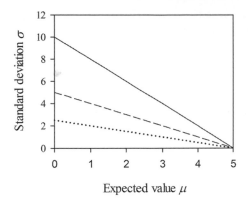

Figure 12.2 Iso-utility lines for the utility function of Eq. (12.6) with $U = 5$, and $k = 0.5$ (solid line), $k = 1$ (dashed line) and $k = 2$ (dotted line).

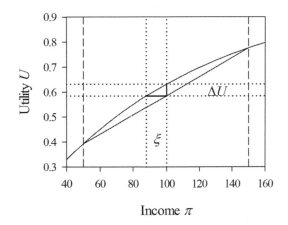

Figure 12.3 Risk-utility function, Eq. (12.7). For details, see text.

these functions represents risk aversion, as will be explained now. Consider the utility function

$$U(\pi) = 1 - e^{-\pi/100} \qquad (12.7)$$

(Fig. 12.3, curved solid line), and assume, as previously, that a decision-maker has the choice between a certain income of $\pi = €100$ and an uncertain income of $\pi_1 = €50$ and $\pi_2 = €150$ with probabilities $p_1 = p_2 = 0.5$ for each outcome (Fig. 12.3, dashed vertical lines).

The expected income is $E\pi = p_1\pi_1 + p_2\pi_2 = €100$ (right vertical dotted line), which equals the certain income in the first alternative, and the associated utility is $U(E\pi) = U(100) = 1 - \exp(-1) \approx 0.63$ (upper horizontal dotted line).

To model risk aversion, one has to swap the sequence of calculations and first map income π into utility, $U(\pi)$, and then build the expected value EU of the obtained utilities. In the first alternative (€100 with certainty) nothing changes and $EU(\pi) = EU(100) = U(100) \approx 0.63$. However, in the lottery the result differs: the two possible outcomes of the lottery yield utilities of $U(50) = 1 - \exp(-0.5) \approx 0.39$ and $U(150) = 1 - \exp(-1.5) \approx 0.78$, respectively, and the expected utility is given by

$$EU(\pi) = p_1 U(\pi_1) + p_2 U(\pi_2) \approx (0.39 + 0.56)/2 \approx 0.58. \qquad (12.8)$$

Graphically, the average $EU(\pi)$ of $U(\pi_1)$ and $U(\pi_2)$ is represented by the solid straight line evaluated at $\pi = 100$, which yields the value represented by the lower horizontal dotted line. It is below the utility $U(E\pi) \approx 0.63$ (upper dotted line) of the first alternative. The geometric reason for this is the curvature of the utility function, which implies that the mean of two different utilities is smaller than the utility of the mean of these utilities. The economic reason is the risk aversion as discussed. To summarise, risk aversion can be modelled by mapping the possible outcomes of a decision into utilities via some concave utility function, and then building the mean of the obtained utilities.

The *certain* income associated with the expected utility of $EU(\pi)$ is termed certainty equivalent CE and defines the certain income that satisfies the decision-maker to the same degree as the uncertain income of the lottery. In the example of Fig. 12.3, the certainty equivalent is given by the equation

$$U(\pi = CE) = EU(\pi) \approx 0.58 \qquad (12.9)$$

which solves for $CE \approx 88$ (left vertical dotted line). The difference

$$\zeta = E\pi - CE \approx 100 - 88 = 12 \qquad (12.10)$$

between the expected income of €100 and the certainty equivalent of 88 is termed the risk premium and compensates for the fact that in the risky alternative the €100 are not earned with certainty but are subject to variability. When deciding to insure, for example, one's belongings, a rational decision-maker should choose an insurance policy so that the demanded insurance premium does not exceed the personal risk premium.

To conclude this outline of decision modelling under risk, several types of risk-utility functions are used in the literature, depending on how risk aversion is defined (Eeckhoudt et al., 2005). Another common function is

$$U(x) = \frac{x^{1-\rho} - 1}{1 - \rho}. \tag{12.11}$$

With these concepts at hand, several of the payment design problems of Chapter 11 can be readily extended to consider risk-averse landowners. Ozanne et al. (2001) analysed their payment design problem under moral hazard (cf. Section 11.2) with the special risk-utility function for the landowners

$$U(\pi) = \pi^{-\theta} \tag{12.12}$$

and found that increasing risk aversion θ in the landowners reduces the optimal monitoring effort and increases the optimal level of environmental protection, because risk-averse landowners are easier to deter from non-compliance by a fine than risk-neutral ones.

The second payment design problem in which risk plays a role is the auction scheme presented in Section 11.4. As derived there, the expected profit of the landowner under certainty equals the difference between bid b_i and conservation cost c_i, multiplied by the subjective probability of the bid being below the bid cap β (Eq. 11.16):

$$\Pi_i = \Pr(b_i < \beta) \cdot (b_i - c_i). \tag{12.13}$$

To consider risk aversion in the landowner, Latacz-Lohmann and van der Hamsvoort (1997) replaced the expected profit Π_i under certainty by the certainty equivalent $CE_i(b_i)$ which is, as described previously, the difference between the expected profit under certainty Π_i and the risk premium ζ, which is a function of the bid: $\zeta = \zeta(b_i)$:

$$CE_i = \Pr(b_i < \beta) \cdot (b_i - c_i - \zeta(b_i)). \tag{12.14}$$

The optimal bid that maximises the certainty equivalent CE_i is then given by (Latacz-Lohmann and van der Hamsvoort 1997)

$$b_i^* = \max\left\{ \beta_{\min}, c_i - \zeta(b_i^*) + \left(1 - \frac{\partial \zeta(b_i)}{\partial b_i}\right)\Bigg|_{b_i = b_i^*} \cdot \left(\frac{\Pr(b_i^* < \beta)}{f(\beta = b_i^*)}\right) \right\} \tag{12.15}$$

with the participation constraint $CE_i \geq 0$. Increasing risk aversion on the one hand reduces the risk premium $\zeta(b_i^*)$ but on the other also reduces its dependence on the bid b_i^*, $1 - \partial\zeta(b_i^*)/ \partial b_i^*$. This, *ceteris paribus*, altogether reduces the optimal bid (ibid.).

The last payment design problem relevant in the context of risk is the agglomeration bonus of Section 11.6. As discussed, the agglomeration bonus leads to a coordination game with multiple (in the example of Table 11.1: two) Nash equilibria. By design of the bonus, the equilibrium associated with spatially agglomerated conservation efforts leads to higher pay-offs for all landowners than the other equilibria. So why, in some of the experimental trials of Parkhurst et al. (2002), did the players end up in one of the other equilibria? Considering the example of Table 11.2, the reason is that a player who chooses to 'cooperate' and conserve the more costly inner land parcel gains only if the other player cooperates, too – but loses otherwise. Since the player does not know whether the other player(s) will cooperate or not, the risk of losing is minimised by choosing the safe bet of conserving the less costly outer land parcel. The more risk-averse a player is, the more likely they will go for a spatially dispersed 'risk-dominant' Nash equilibrium rather than the 'pay-off-dominant' agglomerated Nash equilibrium (Parkhurst et al. 2002, p. 314).

12.3 Fairness and Inequity Aversion

Thus far it has been assumed – as in most economic models – that rational actors are entirely self-interested, so they are only interested in their own material self-benefit and disregard the benefits of all other people. This 'pessimistic' model of human behaviour, however, captures only part of the reality. There are many empirical and experimental examples where people care about the well-being of others or cooperate unconditionally in the sense that they act for the benefit of others even if they cannot expect reciprocity.

Two famous examples may be quoted that demonstrate people's inequity aversion. The first is the ultimatum game (Güth et al. 1982) in which one player receives an amount of money, part of which they are asked to pass on to another player. The second player may accept the offer from the first player and both players get their share. If, however, the second player refuses the offer, both players receive nothing. A rational self-interested player 2 will accept any non-zero offer because it is better than nothing. Knowing that, a rational self-interested player 1 will offer a non-zero share that is as low as possible. In contrast to this expectation, however, in numerous experiments people, when in the position of player 1, offered about 40–50 per cent of their endowment, and when in the position of player 2 rejected very low offers. Obviously, people are averse to income distributions that put them below the average, and punish insultingly low offers – even if it is to their own disadvantage. Knowing that, people in the position of player 1 usually offer almost half of their endowment.

These observations might suggest that people care about equity only as long as they are in the worse or weaker position but act selfishly if they are in the better-off or stronger position. This conclusion, however, still underestimates people's preference for equity, as the second example, the dictator game, demonstrates. The dictator game (Hoffman et al. 1996) is identical to the ultimatum game, except that player 2 has no right to refuse the offer from player 1, so player 1 does not risk their own share even if the share offered to player 2 is too small. Under this setting a purely self-interested rational player 1 should obviously offer nothing. However, in experiments people offered non-zero amounts. These were below those observed in the ultimatum game but significantly positive (Bolton and Ockenfels 2000).

A model to explain these outcomes has been proposed by Fehr and Schmidt (1999), who introduced a utility function that includes inequity aversion. Considering N individuals with incomes π_i ($i = 1, \ldots, N$), the utility U_i of individual i is modelled as

$$U(\pi_1, \ldots, \pi_N) = \pi_i - \frac{\alpha}{N-1} \sum_{j \neq i} \max\{\pi_j - \pi_i, 0\}$$
$$- \frac{\beta}{N-1} \sum_{j \neq i} \max\{\pi_i - \pi_j, 0\} \qquad (12.16)$$

with coefficients α and β obeying $0 \leq \beta < 1$ and $\beta \leq \alpha$. The first sum in Eq. (12.16) considers all individuals with higher profits than individual i and α measures how strongly individual i suffers from their disadvantage; while the second sum considers all individuals with lower profits than individual i and β measures how strongly individual i suffers from their advantage. For the case of two individuals, denoted as 1 and 2, a numerical example is shown in Fig. 12.4.

In Fig. 12.4a one can see that when varying the other individual's profit, one's own utility is maximised if the other individual's profit equals one's own profit, and declines strongly if the other individual's profit exceeds one's own, and less strongly if it falls below one's own

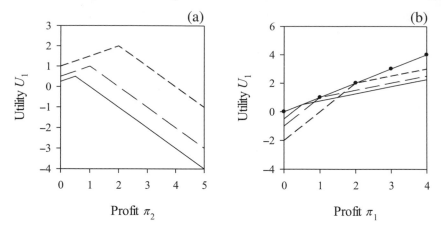

Figure 12.4 Utility U_1 as a function of the other individual's profit π_2 (panel a) and one's own profit π_1 (panel b). Several own profits in panel a: $\pi_1 = 0.5$ (solid line), $\pi_1 = 1$ (long-dashed line) and $\pi_2 = 2$ (short-dashed line); several profits of the other individual in panel b: $\pi_2 = 0.5$ (solid line), $\pi_2 = 1$ (long-dashed line) and $\pi_2 = 2$ (short-dashed line). Lines without circles: $\alpha = 1$, $\beta = 0.5$; line with circles: $\alpha = \beta = 0$.

profit. Fig. 12.4b shows that when varying one's own profit, utility increases strongly as long as the own profit is below the other individual's profit, and less strongly (in particular: less strongly than in the case where *only* one's own profit is considered: $\alpha = \beta = 0$) if it exceeds the other individual's profit. With this model Fehr and Schmidt (1999) were able to reproduce various empirical findings, particularly for the ultimatum and dictator games, quite well.

12.4 Imperfect Information

The availability of information is an important determinant of human decisions. That limited access of the conservation agency to information about the landowners' conservation costs can substantially reduce the cost-effectiveness of a payment scheme has been demonstrated in Section 11.2. In those classical principal–agent models asymmetric information is usually considered such that the principal does not know certain characteristics of the agent while the agent is perfectly informed. As another example, the socially optimal market equilibrium (Section 9.1) can be found only by market participants that are perfectly informed about the other agents' supply costs and willingness to pay.

A recent example in which the agents do not know costs and willingness to pay of other agents is the agent-based water quality trading model by Nguyen et al. (2013). Analogous to emissions trading or the conservation offsets outlined in Section 9.4, in a water quality market water polluters such as farmers and sewage farms are obliged to limit their emissions to a certain threshold. Agents who pollute less earn pollution credits which they can sell to agents who wish to pollute more.

In the model of Nguyen et al. (2013) the agents differ in their abatement costs and in their role in the market: point sources of pollution such as sewage farms act as credit buyers while non-point sources such as farmers act as sellers. Agents of both types act according to strategies that are defined, among others, by the decision whether to participate in the market or not, and − if they are buyers − by the offered credit price and the choice of potential trading partners. Which strategy is most successful depends on the uncertain information about the pollution abatement costs, as well as transaction costs. The transaction costs mainly include the cost of information acquisition to reduce the uncertainty in the estimated abatement costs of other agents. They depend on the set of trading partners found, which in turn depends on the credit price that is demanded.

To solve this problem of imperfect information the authors define a large number of potential strategies and optimise them simultaneously for all agents using a genetic algorithm − an optimisation heuristic similar to the simulated annealing outlined in Section 2.7. To account for the uncertainties, the trading strategies are optimised so that they are as robust as possible to these uncertainties. The optimality criterion in the optimisation is the total abatement cost of all agents.

The most important conclusion from the model analysis is that although the modelled water quality market delivers a higher level of cost-effectiveness (lower total abatement costs) than a regulatory policy under which each agent has to obey their pollution cap, the achieved level of cost-effectiveness of the market is considerably lower than that achieved in an ideal market with perfect information. This is despite the fact that the agents' trading strategies were optimised in a computer-intensive manner. Therefore − as the authors argue − the cost-effectiveness of the simulated market is most likely to be overestimated. As a consequence, cost-effectiveness gains of real-world water quality markets are likely to be far below those predicted by models of ideal markets, and the authors suggest that agencies should assist the agents in the development of their trading strategies to improve market efficiency.

12.5 Bounded Rationality

Tied to the problem of imperfect information is the bounded rationality, that is, the limited cognitive abilities, of the agents (e.g. Rubinstein 1998; Mallard 2016). While Nguyen et al. (2013) point out that the optimisation employed in their study is very likely to overestimate the cognitive abilities of real-world market participants, the pig producers in the cobweb model (Section 2.5) who make the same mistake in every time period by alternatively over- and underestimating the next period's pig price are probably less intelligent than real-world pig producers. As a consequence, prices and quantities in the cobweb model fluctuate wildly, leading to substantial welfare losses, because demand and supply are never in equilibrium and the repeated increase and decrease of pig production leads to transaction costs.

A similar observation of a regular misestimation of demand in a market was made by Drechsler and Hartig (2011) in a model of conservation offsets where conservation costs are uncertain and variable in time and where species habitat cannot be created instantaneously but requires a number of time steps during which habitat is restored. Since in the model credits associated with the creation of habitat are earned only after complete restoration, there is a time lag between investment (restoration effort) and pay-off time (the time when the restoration is complete and the credits are earned and can be sold on the market).

In the model, optimal land-use and trading strategies that show under which conditions one should conserve/restore and when to use the land economically, and under which conditions one should buy or sell credits or stay out of the market, were determined through stochastic dynamic programming (Dixit and Pindyck 1994; Marescot et al. 2013; cf. Section 12.7). To keep the optimisation problem tractable, the agents assume that future conservation costs and the future credit price remain constant at their current levels.

As in the cobweb model, wrong expectations of the agents about future costs and credit prices lead to a repeated oversupply of conservation efforts and credits, associated with efficiency losses. Drechsler and Hartig (2011) speculate that these fluctuations may vanish under the assumptions of smarter agents that predict future conservation costs and credit prices on the basis of observed trends.

While the consideration of smarter agents in the model by Drechsler and Hartig (2011) is beyond the scope of the present section, a first test of the hypothesis might be carried out on the much simpler cobweb model. For this I assume that a pig producer is able to discern whether the

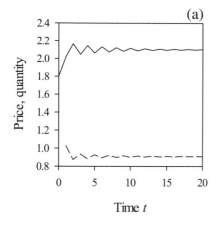
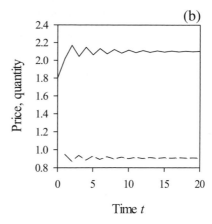

Figure 12.5 Price (solid line) and quantity (dashed line) as functions of time in the modified cobweb model with the supply function of Eq. (12.17). Parameters as in Fig. 2.2: $\alpha = 0$, $\gamma = 2$, and $\beta = 0.48$, $\delta = 0.52$ (panel a) and $\beta = 0.52$, $\delta = 0.48$ (panel b).

market price is currently above or below equilibrium. In periods with comparatively high (low) price the agent predicts the next period's market price as the current price but subtracts (adds) a correction of 10 per cent, so that the first equation of Eq. (2.9) is replaced by

$$s_t = \begin{cases} -\alpha + 0.9\beta p_{t-1} & p_{t-1} \text{ large} \\ -\alpha + 1.1\beta p_{t-1} & p_{t-1} \text{ small} \end{cases}. \qquad (12.17)$$

Figure 12.5 shows that rather independent of the model parameters β and δ, price and quantity converge quite rapidly to their equilibrium values, which demonstrates that market fluctuations may decrease or even disappear if higher cognitive abilities are assumed in the agents.

12.6 Heuristics

In the water quality trading model of Nguyen et al. (2013) the agents have to pay a cost to obtain information about the abatement costs of other agents. The economic benefit of such information to the agent is that they can make better decisions and earn a higher profit. Thus the agent faces a trade-off: pay high information costs to eliminate all uncertainties and maximise one's profit, save the information cost and accept sub-optimal decisions with lower profits, or anything in between by investing moderate information costs and reducing profit losses to a moderate level.

This trade-off in the field of social science was first emphasised by Nobel Laureate Herbert Alexander Simon. Acknowledgement of this trade-off, among others, motivated his proposal on the notion of satisficing behaviour (Simon 1979). An individual with satisficing behaviour does not search the entire decision space to find the optimal decision but searches only until a satisfactory solution is found.

The existence of the trade-off between information cost and the cost of a sub-optimal decision hinges on the (usually plausible) assumption that the marginal profit (as a function of improved information) at some stage becomes smaller than the marginal cost of improved information, because then the net marginal benefit as the difference between profit increase and information cost per added bit of information becomes negative. Only then, in particular if the function of profit versus information is concave (so that the marginal profit declines; cf. Section 12.1) and the function of search cost versus information is convex (so that the marginal information cost increases), is it 'rational' to stop the search process before the 'optimal' decision (that would maximise profit in the absence of information cost) is found and be satisfied with a 'sub-optimal' decision.

The impact of satisficing behaviour on the long-term dynamics of an ecological-economic system has been analysed by Dressler et al. (in press), who modelled pastoralists in drylands who can choose between three types of herd management: 'traditional', where some fraction of pastures is rested depending on environmental conditions; 'short-term profit-maximising', where the size of the herd is maximised by always selecting the pastures with maximum available amount of biomass; and 'satisficing', where the herd size is only maintained above a specified satisfactory level. The model analysis shows that if all pastoralists are profit-maximisers, the number of viable households is higher than if they were all traditional or satisficing pastoralists, but the so-called reserve biomass which ensures long-term availability of edible biomass as well as the associated number of livestock declines substantially in the long run. In contrast, traditional herd management maintains reserve biomass and livestock numbers at constant levels even in the long run; however, the number of viable households is lower than in the case of profit-maximising pastoralists. Satisficing pastoralists are able to achieve some kind of compromise between these two types of outcome: with fewer viable households than under profit-maximising management but stable (though at lower levels than under traditional management) reserve biomass and livestock numbers.

Satisficing behaviour can be observed not only in human agents but also in animals. The Optimal Foraging Theory in Ecology (e.g. Begon et al. 1990, ch. 9.11) describes the foraging behaviour of individuals, assuming that an individual aims at maximising its total food intake for given total foraging time. In this maximisation process the individual considers that foraging time is composed of the time searching for and travelling to a food patch and the time required for collecting food items from that food patch. The more time an individual has spent in a particular food patch and the more food items it has already collected, the more difficult it is to find another food item and the more time is required to find each additional food item. Due to this 'diminishing marginal return' it is not optimal to collect all food items from a patch; it is better to stop the food collection in the current patch at some stage and search and head for another (undepleted) food patch. This stopping rule is equivalent to the human satisficer's decision to end the search for a good decision once a satisfactory decision has been derived.

Next to satisficing behaviour, people often employ other heuristics, that is, protocols to simplify decisions in complex situations or cases of imperfect information. An example of a heuristic is that many people, when confronted with the decision of which wine to order in a restaurant, choose one from a medium price class, possibly because they think the cheap ones are of poor quality while the expensive ones may be overdone or unaffordable to them, or simply because most people feel more comfortable in the middle. If the innkeeper is a rational profit-maximiser and aware of this heuristic, the selected wine will most likely not have the highest quality–price ratio.

As demonstrated in Section 12.5, heuristics can be of differing quality. While the assumption of a constant market price in the cobweb model can lead to substantial fluctuations and is obviously a rather poor heuristic, even a slight improvement in the price expectation as modelled in Eq. (12.17) eliminates the fluctuations and reduces the transaction costs in the pig market. The decisive role of agents' heuristics in the dynamics of a modelled system on the one hand, and the reliance of humans on heuristics on the other hand, makes these an important topic of research in the field of economic and ecological-economic modelling.

12.7 Modelling Change in Behaviour and Decisions

If the context of a decision does not change and a utility function has been defined, the optimal utility-maximising action can be derived on the basis of the available (possibly uncertain) information. Usually, however, systems and decision contexts change over time, and the question is, how are decisions derived under these circumstances and how do they change over time?

Before delving into this issue, one should first be aware of the distinction between actions or measures on the one hand and strategies on the other. An action or measure denotes a certain activity carried out at a particular place at a particular time. A strategy, in contrast, represents a rule set that lays out under which circumstances to carry out what action. Since in dynamic systems these circumstances change over time, strategies will be the focus of this section.

In the context of land use and biodiversity conservation, the simplest strategy, and the closest to static utility maximisation, is to assume that in each time step the optimal land-use measure is determined as in a static setting but under the constraint that its distance from the chosen land-use measure of the previous time step – measured by some predefined metric – stays below some predefined limit. An example is found in Mouysset et al. (2011), who demand that the relative change in land area A_k devoted to some land-use type k, between time steps $t-1$ and t, stays below some limit ε:

$$\frac{|\,A_k(t) - A_k(t-1)\,|}{A_k(t)} \leq \varepsilon. \tag{12.18}$$

The motivation for this restriction is that land-use change involves costs such as purchase or sale of machinery, information acquisition on how to carry out new land use measures and others, so land use should not change too significantly.

A related model for changing land-use measures is by Bulte and Horan (2003) (cf. Section 16.3), who base the rate of change on the relative profitability of different land-use measures. In the model, land of total size L can be used to grow crops on an area A and conserve wildlife on an area $H = L - A$. With π_A and π_H denoting the returns of the two land-use types, the increase in cropping area A is modelled by the equation

$$\frac{dA}{dt} = \varphi(\pi_A - \pi_H), \tag{12.19}$$

so that A increases over time if cropping is currently more profitable than wildlife conservation. Parameter ϕ determines how quickly agricultural area is expanded for a given difference between the two returns π_A and π_H.

If agents interact, the decision context can also change due to the other agent's actions. Examples have been presented in Section 10.3. The strategy tit-for-tat reacts to the behaviour of neighbouring agents, so that cooperation by the other agent in the current time step is responded to by cooperation in the following time step, and defection is responded to by defection.

Given that people often behave according to strategies, one may ask how such a strategy is determined. Two main concepts may be distinguished: optimal control theory and evolutionary approaches. As the terms indicate, the former focuses on the identification of an optimal strategy that maximises some long-term benefit while the latter focuses on the evolution of a strategy in the course of a process. Generally, the evolved strategy will be an optimal one, too, but while optimal control theory guarantees the optimality of the identified strategy, evolutionary approaches cannot provide such a guarantee.

To start with the first approach, in optimal control theory the task is to identify a sequence of actions that maximises a long-term objective. Since the derivation of analytical solutions is often not possible, optimal control problems are often solved through dynamic programming (Marescot et al. 2013), or through stochastic dynamic programming if the system state is affected by stochastic processes. To demonstrate the approach, consider the example by Costello and Polasky (2004), who address a problem of reserve site selection but, in contrast to Ando et al. (1998) in Section 3.1, take dynamics into account. In the analysis, the dynamics arise from the fact that unreserved sites may be lost due to economic development, described by a probability per time period. Furthermore, the budget for purchasing and reserving sites is not fully available in the beginning of the process but only comes in slices every period. Therefore in each period the decision-maker has to decide how many and which reserves to secure through purchase. The decision is complicated by the fact that one and the same species may occur in several sites, so (by definition of the decision problem) a site that contains species already conserved elsewhere adds less to the overall ecological benefit than a site with disjoint species. Thus, the added ecological benefit of a site depends on which sites have already been conserved. In this setting, the task of the conservation manager is to maximise the expected number of protected species at the end of a sequence of T periods.

To cast the dynamic control problem into a mathematical formulation, the authors define a function $V(\mathbf{N}_t, \mathbf{R}_t, \mathbf{B}_t)$ that measures the 'value of the optimal program', defined as the expected number of protected species at the end of period t (and all later periods), given (i) the vector $\mathbf{N}_t = (N_1(t), N_2(t), \ldots, (N_J(t))$ of unreserved (and not yet developed) sites with $N_j(t) = 1$ if site j is unreserved and $N_j(t) = 0$ if it is reserved or has been developed, (ii) the vector $\mathbf{R}_t = (R_1(t), R_2(t), \ldots, (R_J(t))$ of reserved sites with $R_j(t) = 1$ if site j is reserved and $R_j(t) = 0$ otherwise, (iii) the budget B_t available in period t and (iv) that in each period t an optimal decision $\mathbf{X}_t = (X_1(t), X_2(t), \ldots, X_J(t))$ is taken where $X_j(t) = 1$ if site j is purchased in period t and $X_j(t) = 0$ otherwise. The optimality of the decision $\mathbf{X}_t$ is defined by the so-called Bellman equation

$$V(\mathbf{N}_t, \mathbf{R}_t, B_t) = \max_{\mathbf{X}_t \leq \mathbf{N}_t} E_{St} V(\mathbf{N}_{t+1}, \mathbf{R}_{t+1}, B_t), \qquad (12.20)$$

where the max function means that V is maximised over all possible choices of $\mathbf{X}_t$ under the constraint $X_j(t) \leq N_j(t)$ ($j = 1, \ldots, J$) (i.e. only unreserved and undeveloped sites can be purchased), E_{St} is the expectation operator that determines the expected number of conserved species considering all possible futures of the unreserved sites, each of which may become developed ($S_j(t) = 1$) with some probability or stay undeveloped ($S_j(t) = 0$), and $\mathbf{S}(t) = (S_1(t), S_2(t), \ldots, S_J(t))$ is the vector of sites developed in period t. The dynamics of the system are described by the following equations for the number of unreserved and undeveloped sites:

$$\mathbf{N}_{t+1} = \mathbf{N}_t - \mathbf{X}_t - \mathbf{S}_t, \qquad \mathbf{S}_t \leq \mathbf{N}_t - \mathbf{X}_t, \qquad (12.21)$$

the number of reserved sites

$$\mathbf{R}_{t+1} = \mathbf{R}_t + \mathbf{X}_t, \qquad (12.22)$$

and the budget

$$B_{t+1} = \left(B_t + b_t - \sum_j X_j(t) c_j(t) \right) \cdot (1 + \delta), \qquad (12.23)$$

where b_t is the budget slice added to the budget B_t in period t, $c_j(t)$ the acquisition cost for site j in period t, and δ is the interest rate per period that applies if budgets are not fully spent. While money may be saved for later periods, no money may be borrowed, so in each period the budget constraint

$$\sum_j X_j(t) c_j(t) \leq B_t + b_t \qquad (12.24)$$

must be fulfilled.

The solution of the problem is via so-called backward induction: first, the optimal decision $\mathbf{X}_T^*$ for the ultimate period T is determined via Eq. (12.20) as a function of the state variables $\{\mathbf{N}_T,\mathbf{R}_T,B_T\}$ in that period. The associated optimal value, that is, the maximum attainable expected number of conserved species, is $V(\mathbf{N}_T,\mathbf{R}_T,B_T)$. After this optimisation the optimal decision $\mathbf{X}_{T-1}^*$ in the penultimate period $T-1$ is determined, taking into account the possible development of unreserved sizes between periods $T-1$ and T, the implied state variables in the ultimate period, $\mathbf{N}_T,\mathbf{R}_T,B_T$, and the associated value $V(\mathbf{N}_T,\mathbf{R}_T,B_T)$. Thus in the penultimate period it is assumed that in the ultimate period the optimal decision $\mathbf{X}_T^*$ will be taken. Having determined $\mathbf{X}_{T-1}^*$ and the associated value $V(\mathbf{N}_{T-1},\mathbf{R}_{T-1},B_{T-1})$, the optimal decision $\mathbf{X}_{T-2}^*$ in the preceding period $T-2$ is determined. In this manner the optimisation procedure works its way backward in time until the first, present, period is reached.

The authors analyse the described dynamic reserve selection problem for different settings and find that under the optimal sequence of decisions, $\mathbf{X}_1^*$, ..., $\mathbf{X}_T^*$, more species can be conserved than, for example, through a myopic strategy that in each time step considers only the immediate gain in the number of conserved species, ignoring the risk of unreserved sites becoming developed in the future.

While Costello and Polasky (2004) only determined the efficiency gains associated with an optimal sequence of decisions, another useful output of dynamic programming is the derivation of optimal strategies by formulating the optimal decisions (in the example: $\mathbf{X}_t^*$) as functions of the current system state (in the example: $\{\mathbf{N}_T,\mathbf{R}_T,B_T\}$). An example of the explicit derivation of optimal strategies is the previously mentioned analysis by Drechsler and Hartig (2011). In their model, a land-owner using a land parcel for an economic purpose such as agriculture is confronted with the choice between keeping the land parcel in economic use or starting to restore the land parcel – with the option that after the restoration is complete credits are earned which can be sold on the market at some price p; while a landowner currently using a land parcel for conservation can decide between maintaining the land parcel under that use and buying credits at some price p and developing the land into economic use.

What makes the decisions difficult in the setting of Hartig and Drechsler (2011) is that restoration takes several time steps, so it takes time before the restoration efforts pay in terms of earned credits. A long-term investment therefore has to be made whose net reward is uncertain, because the future costs of conservation and restoration (i.e. the forgone profits if the land parcel is not in economic use) vary randomly, and so does the credit price p that emerges from the land-use decisions and the market behaviour of the other landowners. Under the rather simple assumption of each landowner predicting that future conservation costs and credit prices will stay at their current levels, the authors derive three rules for optimal, profit-maximising land use and market behaviour:

1. If the land parcel is currently under conservation and no credits are possessed, the land parcel should be maintained under conservation if the current credit price exceeds the land price of the parcel, where the land price is defined as the sum of all future discounted conservation costs (or equivalently, the sum of all discounted future profits if the land was used for economic purposes). Otherwise, if the credit price is below the land price credits should be purchased and the land parcel turned to economic use.
2. Similarly, if the land parcel is currently conserved and credits are possessed the credits should be sold if the credit price exceeds the land price, while otherwise the credits should be kept and the land parcel developed for economic use.
3. If the land parcel is currently in economic use restoration should be commenced if the benefit of restoration, that is, the discounted price of the credits earned at the end of the restoration process, exceeds the land price. If restoration incurs an additional management cost, the sum of both land price and the discounted management costs must exceed the discounted credit price to make restoration the optimal choice.

As described in Section 10.3, in evolutionary game theory not only the actions or measures can change over time but also the strategies. A core issue in evolutionary game theory is the question of how agents decide whether or not to change their strategy (and also action), and which strategy or action they eventually choose under the given constraints. In Section 10.3 this decision model was introduced as the replicator model (in the literature the term replicator equation is more common, but replicator model is more general and appears more suitable in the context of this book) that uses pay-off as the decision criterion, so that if a neighbouring agent currently has a higher pay-off then their strategy or action is adopted. As demonstrated in Section 10.3, the formulation of the replicator model, in particular the question of whether successful strategies of neighbouring agents are copied or not, can have enormous consequences for the dynamics of the multi-agent system.

A related approach worth mentioning here is reinforcement learning (Roth and Erev 1995; Erev and Roth 1998) that attempts to learn, in an iterative process, a strategy or action that maximises some utility function. The algorithm is based on four principles of the psychology of learning: (i) the tendency to choose a particular strategy or action is reinforced or diminished if the outcome of the action is favourable or unfavourable, respectively; (ii) in the early iterations, learning is faster than in later iterations; (iii) in the exploration of new strategies and actions, new strategies or actions that are similar to the current one are selected (and tested) with higher probability than more different ones and (iv) recent choices and outcomes have a stronger influence on the choice of a new strategy or action than choices that were taken in earlier iterations. Reinforcement learning has been used by Hailu et al. (2005) to model learning in the bidding behaviour of landowners in a conservation auction. An important result from their analysis is that the agents' bidding behaviour converges into a Nash equilibrium (cf. Section 10.2), independent of the agents' initial behaviour.

12.8 Conclusions

As one could expect, the agent-based modelling of human behaviour shows a lot of similarities to the individual-based modelling of plants and animals covered in Section 7.2 (cf. Railsback and Grimm 2012, p. 9 ff.; Iftekhar and Tisdell 2016, p. 539). In both fields, individuals or agents act autonomously and generally attempt to optimise some objective function such as profit or utility in the economic agent-based models, and fitness, reproductive success or survival in the ecological individual-based models. Moreover, information acquisition plays an important role, which in animals involves the sensing of their environment such as stream velocity (van Winkle et al. 1998), hill slope (Pe'er et al. 2013) or pheromone intensity (Watmough and Edelstein-Keshet 1995), and in humans for instance utilities and costs of other agents (Nguyen et al. 2013).

Information acquisition may be costly as in Fahse et al. (1998) where large flocks of birds promise a higher chance of finding food patches but also come with higher food competition with other birds in the flock, and in Nguyen et al. (2013) where it is economically costly to learn the cost functions of other agents. A difference between humans and animals is, of course, the means of information acquisition, which in animals mainly include their senses such as vision and smell, while humans more often use remote sensing such as satellite imagery, telecommunication and the Web. As a consequence, for humans the cost and quality of information should generally decline less strongly with increasing distance than in animals – although Nguyen et al. (2013) provides an example where humans, too, have better information about their close neighbours' situation than about that of further distant agents.

Similarly, memory and future expectations are relevant in both humans and animals. The trout in van Winkle et al. (1998) make predictions about how their behaviour will affect their future growth and mortality, while the agents in the cobweb model and in Drechsler and Hartig (2011) make more or less smart decisions about the long-term profitability of their present land-use choices.

To conclude, there is a lot of similarity between the modelling of human and non-human behaviours. Moreover, such modelling exercises are usually quite complex. The relevant literature is vast, only a tiny fraction of which could be covered in the present chapter. A useful overview on alternative theories of human behaviour and decision-making is provided by Groeneveld et al. (2017) and Schlüter et al. (2017); overviews on agent-based modelling in the fields of economics and land use include, for example, Matthews et al. (2007), Heckbert et al. (2010), Filatova et al (2013) and Hamill and Gilbert (2016).

Recent literature related to Chapter 12 includes Anderies (2000), Hicks et al. (2012), Touza et al. (2013), Epanchin-Niell and Wilen (2015) and Miyasaka et al. (2017).

13 · *The Agglomeration Bonus*

This chapter applies concepts and methods of the Chapters 9–12 to a conservation-economic topic introduced in Section 11.6: the agglomeration bonus (AB). To motivate the analysis I summarise key points of a review by Albers et al. (2018) on the economics of habitat fragmentation. After a brief review of the concept of the AB I present several variants and spin-offs derived from the AB concept which include the application of the AB in conservation auctions and conservation offset schemes, the agglomeration payment, the role of side payments and other instruments for spatial targeting. Although the present book is on modelling, since a large body of literature on the AB analyses the AB's performance experimentally, Section 13.6 provides a brief review of that literature, before the chapter concludes with Section 13.7.

13.1 The Economics of Habitat Fragmentation and the Agglomeration Bonus

Habitat fragmentation has both economic causes and economic consequences, and there is a need for policy instruments to reduce the rate or even halt the process of fragmentation. Although habitat fragmentation can have positive effects on species survival and biodiversity when it provides refugia for less competitive species (cf. Section 8.4) or reduces risks from catastrophes such as fire and diseases (Albers et al. 2018), in general it is perceived as ecologically harmful – not only due to reduced connectivity among habitat patches but also due to an increased proportion of edge habitat which often has lower ecological value (Albers et al. 2018).

Two main policies are discussed in the literature to counteract fragmentation processes: zoning and targeted taxes and payments. Both empirical and theoretical studies indicate that zones can reduce fragmentation pressures (Albers et al. 2018). However, as argued in Section 9.1, regulations such as zoning are likely to be less cost-effective and less accepted among landowners than voluntary, market-based instruments such as conservation payments.

Since homogenous payments are likely to fail to generate agglomerated habitat configurations, Smith and Shogren (2001) proposed an agglomeration bonus that rewards conservation efforts by a bonus payment on top of a homogenous payment if habitat is created adjacent to other habitat. Parkhurst et al. (2002) showed in an experiment that the AB is able to increase the spatial agglomeration of conserved land. In a complex real-world landscape the cost-effectiveness of the AB, however, depends heavily on the initial configuration of conserved land parcels and the employed measure of habitat fragmentation, as Lewis and Plantinga (2007) found in a simulation of land-use dynamics in a region in South Carolina. Nevertheless, the AB concept and its variants have become of increasing interest in conservation policymaking. An example of its practical application, as mentioned by Albers et al. (2018), is the US Conservation Reserve Enhancement Program (CREP) in which 'landowners receive a one-time additional payment when they retire land near other land that is already enrolled in a conservation program'. The following sections present variants and spin-offs derived from the AB concept.

13.2 The Agglomeration Bonus in Conservation Auctions and Conservation Offsets

Although proposed as an extension of a payment scheme, the AB can be incorporated in conservation auctions as well. Here bids for the conservation of land parcels that generate a desired spatial configuration are preferred by the conservation agency to other ecologically less valuable bids. This allows landowners with land parcels adjacent to other conserved land to submit higher bids, which effectively represents an agglomeration bonus (Reeson et al. 2011). Iftekhar and Tisdell (2016) analysed this mechanism, which already has several real-world applications, with the help of an agent-based simulation model. In their analysis the authors also allowed for the formulation of joint bids where a group of landowners with adjacent land parcels submits a bid for that set of land parcels.

The modelled auction was iterated, so that landowners submitted bids, received information about the selection of their bids by the agency, revised their bids and so on. The revision process followed the concept of reinforcement learning (cf. Section 12.7) in which the agents search the strategy space for successful strategies. The reinforcement algorithm by Iftekhar and Tisdell (2016) used two parameters: one that determines how strongly an agent sticks to a previously chosen strategy rather than choosing a new one, and a second one that allows even worse strategies to be selected temporarily – similar to simulated annealing in which the temporary acceptance of worse solutions reduces the risk of getting stuck in a local optimum (Section 2.7).

A major result of the simulation is that without spatial targeting the desired land-use pattern generally will not evolve, while spatial targeting increases the necessary payments. The reason for this is the 'patch selection effect' coined by Drechsler et al. (2010), which describes the fact that to achieve a spatial target, *ceteris paribus*, more costly land parcels have to be selected. The second main result of Iftekhar and Tisdell (2016) concerns the effect of joint bids, which reduces competition among the landowners and thus raises necessary scheme expenses. However, if joint bidding leads to cost synergies among cooperating landowners, it may be financially beneficial for both the conservation agency and landowners.

The AB can, at least in principle, also be applied in conservation offset schemes. Drechsler and Wätzold (2009) simulate a market for conservation credits under changing conservation costs. In each time step landowners have the obligation to conserve a certain proportion of their land but are allowed to trade credits to adapt to the changing conservation costs. Land parcels are arranged on a square grid and each land parcel i can be conserved ($x_i = 1$) or used for economic purposes ($x_i = 0$). Conservation of some land parcel i incurs a cost c_i and earns an amount of credits

$$v_i = 1 + w \sum_{j \in M_i} x_j, \qquad (13.1)$$

where M_i represents the set of land parcels in the Moore neighbourhood (the eight adjacent neighbours) around parcel i. The conservation costs are assumed to be uncorrelated random numbers, which for analytical tractability are drawn in each time step uniformly from the interval $[1 - \sigma, 1 + \sigma]$. The model is a partial-equilibrium model, meaning that other markets or policies are ignored and in each time step supply and demand for credits are in equilibrium. The equilibrium assumption, among others, includes the assumption of an ideal market with a large numbers of market participants without bargaining power.

The model dynamics are controlled by two forces: the randomness in the costs with variation σ which favours spatial dispersion of the conserved land parcels at the locations with lowest conservation costs, and the agglomeration bonus of strength w that favours spatial agglomeration of the conserved land parcels. Like in a ferromagnet (cf. Section 2.1) in which low temperature and/or high magnetic interaction strength g tend to produce an ordered magnetic phase, a large bonus w and/or low cost variation σ leads to an agglomeration of conserved land parcels.

The agents in this model are myopic profit-maximisers in that they base their future land-use decision on the present land-use pattern in their Moore neighbourhood. Hartig and Drechsler (2010) deviate from this assumption and, motivated by Parkhurst et al. (2002), allow the agents to communicate with each other about their intentions on how to use their land in the next time step. This communication leads to higher levels of spatial agglomeration than predicted by Drechsler and Wätzold (2009) and the joint conservation obligation is achieved at lower total cost. The reason for this result is that through communication the agents can search the solution space more effectively to generate a cost-effective land-use pattern.

Another assumption of Drechsler and Wätzold (2009) is that in each time step all land-use decisions occur simultaneously. Relaxing this assumption, Hartig and Drechsler (2010) consider sequential land-use decisions (in random order), which raises the question of who 'owns' the agglomeration bonus. To explain this ambiguity, consider two adjacent land parcels. If initially both land parcels are used economically and the first landowner decides to conserve, they earn credits of an amount 1 (Eq. 13.1). If in the next time step the other landowner decides to conserve, too, they earn credits of an amount 1, and in addition a bonus of magnitude $2w$ is generated: a bonus w for the first conserved land parcel and another w for the second one (Eq. 13.1). The question is: should this bonus of $2w$ be assigned only to the second landowner who effectively generated the spatial agglomeration (option i), or should it be split evenly between both landowners (option ii), since there would be no bonus if the first landowner had not conserved in the first place? An analogous question arises when conserved patches are turned one by one to economic use: who should pay the penalty in the form that they must buy credits? Comparing both options of bonus sharing, Hartig and Drechsler (2010) find that option (i) leads to higher levels of spatial agglomeration and higher levels of cost-effectiveness. The reason is that under option (ii) clusters of spatially connected conserved land parcels 'break up too early' when the cost heterogeneity σ is increased (Hartig and Drechsler 2010).

Both of the studies described here assume that the conservation costs are entirely random, which is quite unlikely to be observed in real agricultural systems. Therefore Hartig and Drechsler (2009) consider temporal and spatio-temporal correlations in the conservation costs. The results show a substantial effect of this correlation on the level of spatial agglomeration and indicate that temporal correlation alone reduces the level of agglomeration while the effect of spatio-temporal correlation appears to be minor.

13.3 The Agglomeration Payment

The agglomeration bonus can be modified in two directions. In the first option, rather than paying the bonus to each individual landowner if their conserved land is adjacent to other conserved land, the conservation agency can address a whole region or group of farmers and offer them a collective payment if they jointly generate a desired agglomerated pattern of conserved land parcels. This approach, which seems simpler and probably associated with lower transaction costs – but is possibly less targeted ecologically – than the individual bonus payments, has various applications in the real world, such as the Swiss network bonus (Krämer and Wätzold 2018).

The second direction in which the agglomeration bonus can be modified is simply to increase the bonus relative to the homogenous base payment (by increasing parameter w in Eq. (13.1)) up to the extreme case in which the base payment is zero and only the bonus is offered. This variant, collective bonus payment and zero base payment, has been considered by Drechsler et al. (2010) and termed agglomeration payment. To operationalise this variant, Drechsler et al. (2010) allowed the landowners in a region to define an arbitrary zone in their region, and in order to earn the agglomeration payment the number (or total area) of conserved land parcels in that zone must exceed a certain target set by the conservation agency.

Since the agglomeration payment is paid only if the conservation target is met, landowners with low conservation costs have an incentive to convince landowners with high costs (and a priori negative profits from participation) to participate in the scheme. To induce the high-cost landowners to participate, the low-cost landowners may offer some of their potential profits in the form of side payments to the high-cost landowners, so that altogether their profits are positive as well. In principle, this incentive also exists in the presence of non-zero base payments but obviously increases with increasing bonus and decreasing base payment.

The consequence of these side payments is that, contrary to homogenous payments (c.f. Section 9.3), the payment does not need to reach the level of the highest conservation cost in the group of landowners but – in an ideal case – only the average cost, because the *ex ante* negative profits of the above-average costly land parcels are fully compensated for by the positive (potential) profits of the below-average costly land parcels. The authors therefore termed this effect the 'surplus transfer effect' through which the landowners essentially pay part of the conservation costs themselves, which reduces the costs incurred by the conservation agency and increases the budget-effectiveness of the agglomeration payment.

In an ecological-economic analysis, Wätzold and Drechsler (2014) compared budget- and cost-effectiveness of the agglomeration payment with those of the agglomeration bonus and with those of homogenous payments and found that the agglomeration payment is never strongly outperformed by the agglomeration bonus and that its relative performance compared with the two alternative payment schemes increases with decreasing conservation budget and decreasing dispersal range of the target species. Much of its comparative advantage is shown to come from the presence of side payments.

This begs the question: do side payments take place at all? Firstly, side payments incur additional transaction costs associated with the negotiations among the landowners, which is likely to reduce the cost-effectiveness of the agglomeration payment. Secondly, even if transaction costs are sufficiently low, the surplus transfer may create distributional problems: would a low-cost landowner really sacrifice some of their (potential) surplus? And would a high-cost landowner accept that offer? To analyse the effect of these distributional issues on the cost-effectiveness of the agglomeration payment, Drechsler (2017c) modelled the landowners as players in an ultimatum game (cf. Section 12.3) that deals exactly with these two questions. While in the absence of inequity aversion low-cost landowners will accept any non-zero offer, in the presence of inequity aversion they will reject insultingly low offers. This bears the risk that the number of participating landowners is too low to reach the conservation target. On the other hand, even though they know that insultingly low offers are rejected, people usually offer much less than their entire endowment (as could be observed in various experiments of the ultimatum game). Consequently, in contrast to the optimistic assumption of Drechsler et al. (2010), only part of the total surplus is actually available for transfer through side payments. Both effects, the higher risk of coordination failure and the lower available surplus, reduce the expected cost-effectiveness of the agglomeration payment in Drechsler (2017c). It could be a matter of future (experimental and field) research to quantify the magnitude of such efficiency losses.

13.4 Side Payments as Means to Improve Compliance

While Drechsler (2017c) analysed the role of side payments with a spatially implicit model (cf. Section 3.1), Bell et al. (2016) developed a spatially differentiated and explicit agent-based model of farmers under an agglomeration bonus scheme. Their model considers bounded-rational utility-maximising farmers that interact with each other through information exchange and side payments.

Each farmer owns several land parcels and for each land parcel has the choice between conventional land use, conservation and cheating (pretending to conserve to collect the payment but applying conventional land use). The land parcels differ in their agricultural productivity. Farmers attempt to maximise expected discounted long-term risk-utility (cf. Sections 12.1 and 12.2). They are bounded rational in that the utility-maximising combination of land-use types on the different land parcels is determined from a limited set of possible combinations, taking only a limited set of possible futures into account.

Each farmer can decide to share information about land-use decisions and pay-offs with others or not. A farmer who participates in this information exchange is able to adopt their land-use according to the received information. In this learning process, information from farmers with a similar farm structure is weighted more strongly than information from farmers with dissimilar farm structure. Each farmer may offer side payments to neighbouring farmers and make educated guesses on whether these will enrol in the payment scheme or not.

The payment scheme consists of a base payment and a bonus for adjacency of conserved land parcels. Cheating is penalised because other farmers are less likely to offer side payments to farmers who have cheated in the past. The main results from the model analysis are that the bonus payment increases participation in the scheme, can decrease overall financial expenses for the scheme and reduces necessary monitoring effort. The reason for the latter result especially is that the bonus, together with the side payments, acts as a diffusion mechanism to spread scheme participation. This in turn is explained by the (realistic) model assumption that farmers offer side payments only to neighbours whom they have observed complying with the scheme previously. In contrast to the conservation agency, farmers do have a good chance of observing cheating neighbours. The risk of being outed as a cheater, together with the negative financial consequences, creates sufficient pressure to comply with the scheme.

13.5 Variants of the Agglomeration Bonus

This section presents two variants of spatial targeting. The first is by Iftekhar and Latacz-Lohmann (2017), who simulate conservation auctions that aim at selecting one out of five possible zones for biodiversity conservation. Each zone contains the land of five landowners who submit bids for conserving their land. Two alternative auction mechanisms are considered: discriminatory price and uniform price (cf. Section 11.4), and for each alternative, several rules for the selection of bids are investigated, such as bid per area (of conservation) and bid per (ecological) value. The performance criteria on which the different auction designs are evaluated are cost-effectiveness, measured by the ratio of ecological benefit and total conservation cost, and budget-effectiveness, measured by the ratio of ecological benefit and total payments.

The model is parameterised for the case of malleefowl (*Leipoa ocellata*) conservation in Australia. The two conservation measures considered are baiting of foxes and reintroduction of malleefowl. The costs of the two conservation measures are estimated from the literature and the benefits are estimated through a simple metapopulation model, also parameterised with data from the literature.

The main result of the model analysis is that the discriminatory price auction generally outperforms the uniform price auction, but that the relative performances depend on the means of bid selection and the level of information available to the bidders.

Another type of spatially targeted auction has been proposed by Polasky et al. (2014). Like the previous instruments, it aims at overcoming the two problems of asymmetric information about conservation costs (cf. Section 11.2) and the spatial interdependence of conserved land parcels with regard to their ecological benefits. In the mechanism proposed by the authors, landowners can submit bids for conserving their land parcel(s). Assuming that these bids equal the conservation costs, the conservation agency determines for each bid b_i the marginal contribution of the associated land parcel i to an objective function that is composed of four terms: the maximised ecological benefits with parcel i *being conserved* and the maximisation taken over all other land parcels $j \neq i$, the associated costs of all land parcels $j \neq i$, the maximised ecological benefits with parcel i *not conserved* and the associated costs of all land parcels $j \neq i$. Adding the two benefits terms and subtracting the two costs terms yields the marginal value ΔW_i of land parcel i, and if that value exceeds the bid b_i landowner i receives a payment equal to that value: $p_i = \Delta W_i$. The authors show that by that choice the generated land-use pattern is cost-effective. The question remains, however, why the conservation agency can assume that the bids equal the conservation costs. As the clue to the mechanism, the authors can show that bidding exactly the true cost is a dominant strategy for each landowner so that deviating from that strategy would reduce their (expected) profit.

A converse variant of the agglomeration bonus is the *agglomeration malus* introduced by Bamière et al. (2013). The authors model the conservation of the little bustard (*Tetrax tetrax*) in a Natura 2000 site in France. Because it is a territorial animal, it is beneficial for the little bustard population if the conserved land parcels are not too close to each other but have direct-neighbour distances between 100 and 1,000 m. Due to spatial correlation of agricultural productivity (and thus conservation costs), a homogenous payment scheme for conservation measures will yield a spatially agglomerated pattern of conservation in those parts of the region where the conservation costs are lowest. To induce spatial dispersion of the conservation measures the authors introduce an agglomeration malus into a homogenous payment scheme which *reduces* a landowner's payment if the conserved land is adjacent to other conserved land.

The agglomeration malus is simulated in a stylised landscape with 900 square land parcels, each of which is 3 ha in size, which can be planted with different crops, some of which are beneficial and some of which are unsuitable for the bird. The agricultural profits and conservation costs associated with the planting of bustard-friendly crops are estimated with a gross-margin calculation based on data from the study region. While under a homogenous payment the simulation generates a spatially agglomerated pattern of bustard habitat, the inclusion of the agglomeration malus delivers the desired distances between the habitats.

A generalised evaluation of agglomeration schemes is provided by Albers et al. (2008), who determine land-use patterns associated with

Nash equilibria. An agent's pay-off is determined by the quantity of conserved land parcels and the number of spatial adjacencies. If spatial agglomeration of conservation efforts is perceived as beneficial by both the conservation agency and landowners, the conservation agency only needs to help in avoiding coordination failure (so that dominated Nash equilibria are avoided: cf. Section 11.6) while if agglomeration is costly for the agents it can be induced only by a non-zero bonus.

13.6 Experiments

Although this book is on modelling, at this point some experimental work should be presented that has been initiated by the agglomeration bonus idea. Next to Parkhurst et al. (2002), the first was Parkhurst and Shogren (2007) who carried out an experiment with four players, each owning a quarter of a square grid with altogether 10 by 10 land parcels. The players were told their pay-offs were dependent on their land-use choices and they were allowed to communicate with each other. Different levels of agglomeration bonuses for north–south and east–west borders were offered to the players that effectively were set to induce conservation of a core, a cross, a corridor or four corners in the model landscape, respectively. Each scenario was played for 30 rounds. The players turned out to be able to generate the desired land-use patterns on average in about 70–90 per cent of all trials.

Another set of experiments was carried out by Banerjee et al. (2012, 2014), who arranged the players in a circle so that each player had two direct neighbours. Taking up the question of Parkhurst et al. (2002) whether players would play the pay-off-dominant agglomerated strategy or the risk-dominant dispersed strategy, Banerjee et al. (2012) found that the frequency of the pay-off-dominant strategy declines with increasing number of players (12 versus six). While in their paper each player was informed only by their direct neighbours' present and past land-use decision, in Banerjee et al. (2014) the information space was expanded to the two players next to the two direct neighbours. In these experiments the increased information space led to a higher level of coordination (agglomeration) in the first rounds, which however declined with more rounds being played.

An alternative approach to spatial targeting was taken by Kuhfuss et al. (2016) who carried out a choice experiment (cf. Section 16.5) with wine growers in the South of France to determine their willingness to accept a collective bonus (a group payment similar to the Swiss network bonus analysed by Krämer and Wätzold (2018)) for the joint conservation of land in a region. The authors found that the collective bonus increased the level of participation in the scheme and reduced the cost per hectare of conserved land.

While economic research has traditionally focused on financial incentives, a recent approach to environmental policy is to utilise nudges. Nudges represent behavioural incentives and often employ social norms, that is, expected patterns of behaviour and belief. In an experiment by Kuhfuss et al. (2018) the spatially agglomerated conservation of land was incentivised by either a financial bonus or a nudge. The nudge was that the joint biodiversity benefit from the generated land-use pattern in the group was measured and compared with the benefits produced by other groups in the experiment. The ranking of the benefits was communicated to each group, so well-performing and poorly performing groups were informed of their high and low performances, respectively. Although the nudge was found to be less effective than the agglomeration bonus, its effect on the level of conservation and spatial agglomeration in the group was significant.

As described previously, the agglomeration bonus is also applied in conservation auctions. Using the same circular design as outlined previously, Banerjee et al. (2015) investigated the effect of information on the behaviour of the players and the cost-effectiveness of the scheme. As shown by the previous experiments, information is likely to improve the spatial coordination of conservation efforts, which should also apply in auctions. On the other hand, the ranking of one's own bid by the conservation agency may increase rent seeking in that landowners with highly ranked bids become aware of their bargaining power and raise their bids in future rounds. This effect was indeed observed by the authors. Rent seeking has been found to be particularly large if landowners are allowed to submit joint bids, where two landowners with adjacent land parcels may submit a single bid for the two land parcels (Banerjee et al. 2018).

There are, however, ways to reduce rent seeking. According to the findings of Reeson et al. (2011, 2018), these include keeping the number of auction rounds unknown to the landowners, restricting the revision of bids between rounds and making the levels of all bids (not the scores as in Banerjee et al. (2015)) known to all landowners in the auction.

Thus far there seems to be only a single experiment with the application of the agglomeration bonus to conservation offset schemes, which is by Parkhurst et al. (2016). In contrast to the model by Drechsler and

Wätzold (2009), the bonus in the experiment was not included in the trading rules, but the credits market itself was non-spatial while the bonus was added as an extra incentive for landowners who conserved adjacent land parcels. As in the experiments with conservation payments, the players were informed about the pay-offs of their land-use choices and allowed to communicate with each other. In the experiment the conservation offset scheme alone reduced the overall cost of conservation without delivering any spatial agglomeration, while the inclusion of the agglomeration bonus increased the overall cost but also led to a higher level of agglomeration.

13.7 Conclusions

To conclude this chapter, the agglomeration bonus has enormous potential to improve both the level and the spatial coordination of conservation efforts. Evidence comes from both theoretical and experimental work and applies to conservation payments, auctions and conservation offsets.

Major challenges in the implementation of the agglomeration bonus include (i) the trade-off between pay-off dominance (where conservation efforts are spatially agglomerated and individual pay-offs are higher, but the risk of losing is also higher if other landowners do not join) and risk dominance (where conservation efforts are spatially dispersed, and where both the pay-off and the risk of losing are lower); and (ii) asymmetric information and rent seeking of landowners.

These challenges can be solved to some extent by an appropriate scheme design which involves, among others, the appropriate type and amount of information provided by the conservation agency to the landowners and the facilitation of communication among the landowners. A further issue that should not be overlooked is social norms that affect the behaviour of the landowners beyond financial incentives.

Part IV
Ecological–Economic Modelling

14 · *Foundations of Ecological-Economic Modelling*

The two previous Parts II and III provided an overview on the fields of ecological and economic modelling. As demonstrated in Part I, modelling is an essential research approach in many scientific disciplines. We saw that models can have many different purposes. In particular, they complement empirical and experimental work by mediating between theory and the real world (Morrison and Morgan 1999a). On the one hand, models shape our thinking and help us develop experiments to learn more about the natural and social systems around us. On the other hand, empirical and experimental knowledge can improve our models.

What, then, is the discipline to provide a 'home' for ecological-economic modellers? To answer this, one should realise that the ecological models presented in Part II, which focus on the dynamics of species populations in spatially structured and stochastic environments – with the aim to understand the drivers of biodiversity – only form a subset of all ecological models. These drivers of biodiversity include, in particular, human land use, which affects among others the spatial distribution of habitat for the species. Many of the models in Part II therefore belong to the field of conservation biology, which is a subset of ecology.

Very similarly, the models presented in Part III are only a small subset of economic models, those that focus on the economic reasons for environmental degradation (in the example of biodiversity loss) and economic instruments for the cost–effective and efficient protection of the environment. The concepts outlined in Parts II and III together encompass much of the research field of biodiversity economics, which may be viewed as an extension of conservation biology in the economic dimension, or an extension of economics in the ecological dimension. Biodiversity economics is currently becoming established as a distinct research field, as indicated for example by textbooks such as Kontoleon et al. (2007) and the research and policy project TEEB (The Economics of Environment and Biodiversity: cf. Section 16.7). To answer the question posed earlier: the disciplinary home of ecological–economic modellers is biodiversity economics.

Two different economic disciplines claim to include biodiversity economics as one of their research fields: environmental economics and ecological economics. To understand the conceptual fundaments of ecological–economic modelling it is necessary to learn some of the basic concepts of these two disciplines. In the following I will (very) briefly outline the history of environmental–economic and ecological–economic thought and the differences and similarities between environmental and ecological economics. This will finally feed into implications and recommendations for ecological–economic modelling.

14.1 The History of Environmental and Ecological–Economic Thought in a Nutshell

This abstract of the history of economic thought is based on Edwards-Jones et al. (2000, ch. 2) and starts with the father of modern economics, Adam Smith (1723–90). Smith argued that in a perfect market economy the self-interested behaviour of individuals to maximise their own well-being implies the maximisation of the well-being of the society as a whole. A contemporary of the Industrial Revolution in Great Britain, Smith assumed there was no shortage (no 'absolute scarcity': see Section 14.3) of natural resources (agricultural land) that could impede the production of goods and the functioning of the market.

Thomas Malthus (1766–1834) experienced more of the downside of the Industrial Revolution and was more pessimistic about the future of society. Assuming an exponential growth of the human population but only a linear growth in food production due to decreasing marginal productivity, he predicted a shortage of food and either starvation or involuntary control of the human population. Although Malthus was overly pessimistic, ignoring, for example, technological progress, his general concern about the limits to growth is still relevant, especially in our times.

The economist and philosopher Karl Marx (1818–83) was one of the first writers to emphasise the importance of justice and equity in economic reasoning. In his era there was a large gap between the capitalist class, who owned most of the resources and production infrastructure, and the working class. In agreement with Adam Smith and contrary to Thomas Malthus, in Marx's view there were enough resources available to meet the needs of all individuals, but the resources were misallocated. In addition, Marx assumed that the observed environmental degradation and overuse of resources was simply a consequence of the capitalist economy and could, together with social injustice, be remedied by the establishment of a communist economy.

At the end of the nineteenth century, the neoclassical school of economics turned the focus away from the production side (as emphasised by Marx) to the demand side and introduced the concept of utility, including that of diminishing marginal utility (cf. Sections 9.1 and 12.1). The neoclassicists argued that it is best to leave decisions of resource and goods allocation to the individuals in the market who were assumed to decide rationally, and to be perfectly informed and self-interested. The important signals for resource use and goods production as well as for investments in technological innovation are the prices formed by the market. Thus, economic growth could again be sustained even in a world of limited resources.

With the economic problem of efficiency being 'solved', the problem of equity was addressed by Vinfredo Pareto (1848–1923). Pareto introduced the concept of social welfare, which in a way is the sum of all members' net utilities. In Pareto's view an economy is efficient if no person can be made better off without making someone else worse off. This efficient allocation is guaranteed by the market. The problem of inequity can be solved rather easily through economic growth so that overall wealth increases and the surplus can be allocated to the worse-off without taking anything away from the better-off.

This optimistic view of a market that solves all problems of scarcity, inefficiency and injustice was challenged in the early twentieth century by an increasing awareness of overpopulation, resource shortages and external effects of human activities on the environment. Externalities at that time were most visible in the form of pollution, which on the one hand allowed for cheap goods production, maximising the net benefits of both producers and consumers, but on the other hand reduced the welfare of the society as a whole.

A solution to pollution (and other problems of environmental degradation) was proposed by Arthur Pigou: a tax that is set exactly at a level so that the total marginal benefits and total marginal costs of pollution to society are equal. Assuming profit-maximising behaviour, the polluters will pollute exactly at the level at which the marginal costs equal the Pigouvian tax – which eventually implies that the level of pollution is efficient, that is, it maximises social welfare (cf. Section 9.3).

Another proposed solution was to put public goods such as clean water under private ownership. Ronald Coase showed that if property rights are clearly defined, then those people benefiting from pollution and those being harmed by it will, through negotiation, agree on an efficient level of pollution at which the marginal benefit and the marginal cost of pollution are equal (for the practicality of this approach, see Edwards-Jones et al. (2000, chs 12 and 14)). Although nowadays the environment is still largely a public good, Coase's ideas have had an important influence on the development of market-based instruments for environmental protection such as emissions trading.

Possibly for reasons presented in Sections 14.2 and 14.3, the concepts proposed by Pigou and Coase failed to convince a number of economists, such as Kenneth Boulding who argued that earth is (largely) a closed system and unbounded economic growth is not feasible. His solution was a circular and efficient economy that reuses materials as much as possible and utilises resources as efficiently as possible. Boulding's concerns were cast into a mathematical model by Dennis Meadows and colleagues (1972) which predicted a continuing increase in resource use and pollution, followed by a collapse of human population size and well-being.

Although Meadows' model (similar to that of Malthus) was overly pessimistic, it may have been a cause for the development of the discipline of ecological economics, which doubts the power of markets to solve all problems of environmental degradation and overuse of resources. Instead, the embeddedness of the economy within the biophysical world was stressed, together with the dependence of humans on natural resources and a healthy environment and the tight and two-way interactions between humans and their natural environment.

A second spark for the rise of ecological economics was certainly the Brundtland report published in 1987 by the World Commission on Environment and Development (WCED). A core theme in this report is sustainable development, which is defined as 'development that meets the needs of the present without compromising the ability of future generations to meet their own needs' (WCED 1987). In this definition, sustainable development implies non-declining welfare and consumption over time and non-declining capital stocks over time, where capital includes natural, human-made (e.g. infrastructure), human (e.g. education) and social (e.g. norms and freedom) capital. Sustainability and the interaction between the economy and the biophysical world may be regarded as the two core issues of ecological economics (Baumgärtner et al. 2008). These issues are now widely accepted even beyond the community of ecological economists, while the potential of markets as important instruments for environmental protection is acknowledged by many ecological economists. More details about the 'philosophies' of environmental and ecological economics and their further evolution to the present time are presented in Sections 14.2 and 14.3.

14.2 Environmental Economics in a Nutshell

The realisation that economic markets may fail to solve environmental problems and that additional instruments, such as the designation of property rights over public goods (Coase) or the taxation of environmentally harmful behaviour (Pigou), are required may be regarded as the origin of environmental economics (Hanley et al. 2007; Haab and Whitehead 2014). Assuming rational and self-interested behaviour of the individuals and still relying on the concept of an economic market, environmental economics may be regarded as an application or extension of neoclassical economics (Costanza 1989).

The core theme of environmental economics, borrowed from neoclassical economics, is the maximisation of ends (utility or welfare) by scarce means (resources) that can have alternative uses (provision of different goods and services) (Baumgärtner 2006). In an economy, the instrument of welfare maximisation is the market, which for perfect functioning relies on a number of preconditions (Hanley et al. 2007, ch. 3.2):

1. property rights must be well-defined (comprehensive, exclusive, secure and transferrable) so that market participants can freely exchange goods and services;
2. producers and consumers behave competitively by maximising their benefits and minimising their costs;
3. market prices are known by producers and consumers; and
4. transaction costs, such as the cost to acquire information and organise the market, are zero.

While precondition (i) relates to the previously discussed ideas of Ronald Coase, preconditions (ii) and (iii) reflect the classical image economics has about humans, termed the *homo oeconomicus*: that they are perfectly informed about the (present and future) benefits and costs of their actions and that they strive, based on that perfect information, to maximise their individual utility. The correspondence to Adam Smith's early ideas about the equivalence of the individual's and society's interests is obvious.

As already demonstrated in Section 9.1, markets often fail in the efficient protection of the environment and the efficient allocation of natural resources. Major reasons for this are the characteristics of the environment as a public good (non-excludability and non-rivalry, which leads to understated preferences for these goods) and the presence of external effects (one person creates benefits or costs to another person without receiving, or paying, compensation) − both of which may be regarded as two sides of the same coin (Fees and Seliger 2013, ch. 3.3). Other possible causes of market failure are asymmetric information (cf. Chapter 10.2) and non-convexities (Hanley et al. 2007, ch. 3.6). Non-convexities mean that, contrary to the usual assumption (cf. Fig. 9.1), benefit functions are not concave and/or cost functions are not convex. Mathematically, the problematic consequence of non-concave benefits and non-convex costs is that here the first-order criterion for efficiency, the equality between marginal benefits and marginal costs (Eq. 9.2), is only necessary for efficiency but not sufficient.

With the Pigouvian tax being the first example, a central approach of environmental economics for the remediation of market failure and the solution of environmental problems is market-based instruments such as the conservation payments and conservation offsets introduced in Sections 9.3 and 9.4.

To conclude, the mainstream in the discipline of environmental economics focuses on optimisation (of utility or welfare), steady states or equilibria (as in markets), and efficiency and cost-effectiveness (with lesser focus on equity and distribution of wealth). Social utility or welfare is assumed to be the sum of the individuals' utilities, and in the maximisation of their utility individuals can freely substitute resources with other resources while basic needs for survival are assumed to be always guaranteed. In addition, the linkage of the economy to the biophysical world is often considered in a rather simplistic manner, and although biodiversity is addressed, as in forest economics and fishery economics, it is often considered as a simple scalar stock that does not interact with other species nor with the ecosystem as a whole.

14.3 Ecological Economics in a Nutshell

Despite its intriguing simplicity and logic, environmental economics began to receive some criticism in the 1970s and 1980s from a number of economists and natural scientists, such as Herman Daly, John Proops and Robert Costanza, who proposed a new economic discipline:

ecological economics (Edwards-Jones et al. 2000; Daly and Farley 2010; Martinez-Allier and Muradian 2015). Their criticism of environmental economics may be sorted along two dimensions: a system-theoretic one and another one concerned with value and justice.

Central to the system-theoretic critique is the argument that environmental economics, which is rooted in neoclassical economics, ignores the biophysical limits around economies and has a too simplistic view of the interaction between the economy and the biophysical world. Environmental economics was thus established as a new discipline to study, among others, how ecological and economic systems interrelate and co-evolve (Proops 1989), and to establish a scientific discipline that includes more economics than the existing ecological disciplines and more ecology than the existing economic disciplines (Costanza 1989).

Taking a more system-theoretic perspective, ecological economics absorbed many of the new concepts being developed at the time, including non-equilibrium thermodynamics, self-organisation, feedbacks, non-linearities, irreversibility and resilience (Edwards-Jones et al. 2000, Røpke 2005). Resilience was coined by Holling (1973) as the ability of an (eco-) system to return to its initial domain after it has been disturbed by some external influence. By the somewhat vague term 'domain' I mean the system's typical organisational structure (e.g. abundance of and interactions between system components such as species populations in an ecosystem), as well as its typical spatial structure and temporal evolution. Simple examples of 'domains' are the magnetic and non-magnetic phases of a ferromagnet (Section 2.1), a savannah with its coexistence of grass and trees (Section 8.3) or the two phases of spatially clustered and scattered habitat arrangements produced by a conservation offset scheme with an agglomeration bonus (Section 13.2).

The second dimension of the critique of the discipline of environmental economics is its neglect of equity and justice (Costanza et al. 1997, Faber 2008). Edwards-Jones et al. (2000, ch. 5.3) distinguish three types of justice in an economy: commutative justice, which includes, for example, fair prices that are not dictated by someone with greater power; productive justice, which includes, for example, participation in the market and satisfaction of one's own needs; and distributive justice, which includes the fair distribution of resources and goods. Markets cannot guarantee justice, because they 'reflect the predominant power relationships and existing institutional arrangements' (Røpke 2005, p. 278), which are unlikely to have evolved with the aim of establishing a state of justice.

Making just decisions involves the prioritisation of values, which form an essential theme in ecological economics. Values may be direct and marketable (e.g. in the case of timber from a forest), direct and non-marketable (e.g. landscape scenery), indirect (e.g. climate regulation), of an option type (e.g. a plant that may in the future turn out to contain a valuable medical substance) and intrinsic (e.g. the life of an organism per se) (Edwards-Jones et al. 2000, ch. 6.3). Since environmental economics assumes perfect substitutability, all values can be measured on a single, monetary, scale, which is challenged by ecological economists who argue that resources or goods are not always substitutable and in those cases cannot or should not be monetised.

A reason for limited substitutability is absolute scarcity (Baumgärtner 2006; Hanley et al. 2007, ch. 1.2), which acknowledges that a certain amount of some resources such as food, (clean) water and (clean) air is indispensable for survival. Whatever their market prices, a starving individual will always 'choose the bread instead of the compact disc' (Baumgärtner 2006). Ignoring resource limitations, environmental economics cannot address the issue of absolute scarcity but focuses on relative scarcity, which occurs because the use of a unit of a resource for one purpose prevents the use of that unit for another purpose. In that context an individual has the free choice to allocate its (scarce) resources to maximise its utility, which via the (ideal) market also maximises social welfare.

Because it includes the dynamic interactions within and between ecological and economic systems and acknowledges justice as a primary objective of policymaking, ecological economics appears to be well suited to address the problem of sustainability (cf. Section 14.1). This is due to the nature of the concept which demands a just allocation of resources among the present and all future generations. A second reason for the discipline's ability to address sustainability issues is its previously mentioned emphasis on resilience. The two concepts of sustainability and resilience are related in that both deal with the question whether a system can be maintained within certain limits (capital stocks in the case of sustainability) (cf. Common and Perrings 1992; Derissen et al. 2011). The main difference between the two concepts is that resilience is a value-free concept that does not judge whether the transition from one domain to another is desirable or not, while the concept of sustainability involves value judgements about the relative preferabilities of the different domains.

Another difference between environmental economics and ecological economics is the prevalent definition of sustainability: since within environmental economics substitutability is assumed to be unlimited, a development is sustainable if the *aggregate* of all capital stocks does not decline over time (so-called weak sustainability); if, in contrast, substitutability is limited as acknowledged in ecological economics, development can only be sustainable if *none* of the different capital stocks – especially natural capital – declines over time (so-called strong sustainability) (Edwards-Jones et al. 2000, ch. 2.8; Hanley et al. 2007, ch. 2).

Two further features or requirements of ecological economics should be mentioned: transdisciplinarity and pluralism. Transdisciplinarity means that ecological-economic research should involve not only scientists but also people outside the scientific community including politicians and 'ordinary' citizens: producers and consumers of goods and services, people benefiting from resource use and those harmed by it, and others (Baumgärtner et al. 2008). The reason for this requirement stems from the complexity and severity of many economic–environmental problems; the fact that a multitude of people are involved and affected; and that on the one hand these people can often provide valuable information, while on the other hand their inclusion raises the chance that scientific recommendations are adopted in practice.

Pluralism is a consequence of the development of the discipline that was founded to broaden the view of environmental problems beyond that of environmental economics and consider all relevant factors of a studied problem and their interactions. This requires the inclusion of a multitude of disciplines together with their concepts, methods and abilities, since there is 'probably no one *right* approach or paradigm' (Costanza 1989, p. 2). The advantage of pluralism is that it insures against an overly narrow view of environmental problems and their possible solutions; a disadvantage might be that the discipline loses its distinctiveness and develops into a state of arbitrariness. This may aggravate the problem that the knowledge of ecological economics is less well structured and organised than that of environmental economics, for example, which explains why ecological economists have a less unified discourse than environmental economists and may explain why currently the discipline is drifting towards environmental economics (Plumecocq 2014).

14.4 Implications for Ecological–Economic Modelling

Modelling is an important research approach in the field of ecological economics (Proops and Safonov 2004). A first agenda for modelling research for the sustainable management of ecological–economic systems (which of course includes the field of biodiversity economics) has been formulated by Costanza et al. (1993, p. 553):

Application of the evolutionary paradigm to modelling ecological economic systems. The evolutionary paradigm provides a general framework for complex ecological economic systems dynamics. It incorporates the elements of uncertainty, surprise, learning, path dependence, multiple equilibria, suboptimal performance, lock-in, and thermodynamic constraints. . . . Key methods include adaptive computer simulation models and integrated conventional/evolutionary game theory.

Scale and hierarchy considerations in modelling ecological economic systems. The key questions involve exactly how hierarchical levels interact with each other and how to further develop the three basic methods of scaling (statistical expectations, partitioning, and recalibration) for application to complex ecological economic systems. . . .

The nature and limits of predictability in modelling ecological economic systems. The significant effects of nonlinearities raise some interesting questions about the influence of resolution (including spatial, temporal, and component) on the performance of models, in particular on their predictability.

With regard to sustainability research, a helpful modelling concept has been proposed by Baumgärtner and Quaas (2009). Given the previously mentioned limits of predictability – whether due to non-linear dynamics and feedback loops or due to stochasticity – minimum thresholds on capital stocks, utilities, resources or other system variables usually cannot be maintained with absolute certainty but only with a certain probability. A system is 'ecologically-economically viable', according to Baumgärtner and Quaas (2009), if specified thresholds for the specified relevant stocks and services are not undercut with higher probabilities than are specified. This concept can be regarded as an extension of the population viability concept introduced in Section 5.3 and has been applied, for example, by Mouysset et al. (2013) to the farmland bird conservation problem of Mouysset et al. (2011). Derissen et al. (2011) analysed, based on a fictitious ecological-economic system, under what conditions sustainability is necessary and/or sufficient for resilience.

Although markets are not explicitly mentioned in the approaches and recommendations discussed here, market-based instruments for environmental protection and biodiversity conservation are a rewarding research topic for ecological-economic modelling. Despite valid criticism, their role in environmental protection is significant. Additionally, although the rational, perfectly informed and self-interested *homo oeconomicus* is, of course, a caricature of human behaviour, it is able to explain much of the observed behaviours of humans and provides a good starting point from which to add more realistic human behaviour (cf. Sections 12.3, 12.5 and 12.6). Despite its limitations, the advantage of the model of the *homo oeconomicus* and its variants is the simplicity with which it can be operationalised in mathematical models – which might explain why it is the most widely used model of human behaviour (Schlüter et al. 2017).

Having reviewed the main concepts of ecological economics and environmental economics, one can conclude that there is a substantial overlap between the two. While Martínez-Allier and Muradian (2015), for instance, acknowledge market-based instruments such as conservation payments as important topics of ecological-economic research, sustainability and resilience as well as other 'originally ecological-economic' concepts are discussed in textbooks on environmental economics such as Hanley et al. (2007) and Tietenberg and Lewis (2018). In the development of ecological-economic models one should make use of the useful concepts of both disciplines, always keeping in mind that it is not its attachment to a scientific discipline that makes a good model but its ability to fulfil its purpose.

15 · *Benefits and Challenges of Ecological-Economic Modelling*

15.1 Why Is Ecological-Economic Modelling Useful?

A first quick answer to the question of why ecological-economic modelling is useful might be that modelling is an integral part of many disciplines: many prominent physicists are or were theoretical physicists who model physical phenomena to complement the research of experimental physicists; ecological modellers are less abundant than field ecologists but make valuable contributions to the field of ecology; and economic models are taught from the very beginning of an economics curriculum at most universities. The interdisciplinary field of biodiversity economics mentioned in Chapter 14 has a modelling community, too – which is doing ecological-economic modelling.

For a more detailed answer to the question why ecological-economic modelling is useful, one may recall that ecological modellers usually only investigate ecological questions such as the survival of species (cf. Chapter 5) or the conditions facilitating biodiversity (cf. Chapter 8). However, the amount and spatial distribution of habitat for these species often depends on human land use, and even natural disturbances such as fires and floods that have impacts on species coexistence are often induced or at least altered by humans. This socioeconomic dimension to biodiversity is rarely addressed in ecological models. It is, in contrast, addressed by (environmental-) economic models that analyse, for example, how an incentive mechanism such as the agglomeration bonus (cf. Chapter 13) affects the amount and spatial arrangement of habitats. However, these economic model analyses often do not go any further to evaluate how the induced habitat pattern eventually affects the survival of species.

Despite the valuable insights drawn from disciplinary ecological and economic models, various important questions can be analysed and answered only by linking the ecological and economic dimensions (cf. the Editorial in *Conservation Biology* by Teel et al. (2017)). Relevant topics here are the cost-effective allocation of conservation efforts

Figure 15.1 Fictitious landscape composed of six land parcels, differing in the number of endangered species they harbour and their conservation cost (after F. Wätzold, pers. comm.).

(e.g. Ando et al. 1998; see the discussion of Fig. 15.1) and the design of cost-effective economic instruments for the conservation of species. Other important research questions that require the integrated consideration of ecology and economics (and often even additional scientific disciplines) are sustainability and resilience (cf. Section 14.3), which can to some extent be addressed through disciplinary research, but drivers and feedbacks to and from other dimensions of the studied system are usually ignored.

Having determined that integrated ecological–economic research often leads to a better understanding of ecosystem dynamics and better recommendations for the sustainable management of these systems, the question is how this integration should be organised. Sometimes, especially in the early years of interdisciplinary environmental research, a system of interest is investigated separately by researchers from different disciplines, and after each discipline has formulated some management recommendations, a joint list of policy conclusions is compiled. This approach may work in some instances but must fail in at least two types of conservation problems (Wätzold et al. 2006):

1. when management solutions are incompatible between disciplines, and
2. when the system is governed by feedback loops.

The first issue of incompatible solutions may be demonstrated by a simple example from lecture notes by Frank Wätzold. Consider a set of six land parcels. each of which can be conserved or used for economic purposes (Fig. 15.1). Conservation of parcel i incurs a cost (e.g. forgone agricultural profit) C_i and yields an ecological benefit (e.g. number of species conserved) B_i. For simplicity assume that, like the costs, the ecological benefits B_i are additive over conserved land parcels (which is the case, e.g., if the species sets in the different land parcels are disjoint and all species are equally valuable), so that the total ecological benefit B_{tot} is the sum over the B_i of all conserved land parcels.

Confronted with this setting, consider a purely 'ecological' decision-maker who is asked to select one or more land parcels for conservation. Ignoring the costs C_i, they will first select the ecologically most valuable one with $B_i = 6$, and if they are allowed to select further land parcels, this land parcel will be accompanied by the parcels with $B_i = 5$, $B_i = 4$ and so on. Thus, the 'ecological' decision-maker will maximise the total ecological benefit B_{tot} for a given number of selected land parcels. Now consider an 'economic' decision-maker. Being, admittedly, a little naïve, they will first select the cheapest land parcel at a cost of $C_i = €1,000$, and if allowed to select more than one, they will add one or both of the other two parcels with $C_i = €1,000$, and then the next expensive ones. In this way the 'economic' decision-maker minimises the total cost C_{tot} for a given number of conserved land parcels.

In the third consideration of the conservation problem let the task be to maximise the total ecological benefit B_{tot} for a given total cost of $C_{tot} = €3,000$. Based on the decision rules modelled previously, the 'ecological' decision-maker will select the upper left land parcel in Fig. 15.1, which yields an ecological benefit of $B_{tot} = B_1 = 6$. The 'economic' decision-maker will select the three lower right parcels with total ecological benefit $B_{tot} = B_4 + B_5 + B_6 = 6$. Another selection, however, will lead to a larger total ecological benefit: conservation of the two land parcels $i = 2$ and $i = 3$ meets the cost constraint but delivers a total ecological benefit of $B_{tot} = B_2 + B_3 = 9$.

This simple example contains two messages. Firstly, by applying an integrated consideration of ecology (benefits B_i) and economy (costs C_i), a higher level of cost-effectiveness can be attained than under the pure disciplinary analyses. Secondly, the cost-effective set of land parcels, $I = \{2,3\}$, is contained neither in the 'ecological' set, $I = \{1\}$, nor in the 'economic' set, $I = \{4,5,6\}$. Thus, no aggregation procedure of the two disciplinary solutions can be formulated that delivers the cost-effective solution.

Are these conclusions, drawn from a fictitious toy example, transferrable to real-world conservation problems? The first conclusion, that the integrated consideration of ecology and economy can lead to substantial cost-effectiveness gains, has been demonstrated in the seminal paper by Ando et al. (1998) outlined in Section 3.1. Furthermore, figure 2 of that paper shows in colour the counties of the United States that should be conserved (note that only one acre in each county is considered for conservation): yellow for the 'ecological' selection that minimises the number of conserved counties required to protect all species, blue for the cost-effective selection and green for those counties that appear in both selections. The number of green counties is rather small, so there is little overlap between the 'ecological' and the cost-effective selections. Although the authors did not show the 'economic' selection that maximises the number of conserved counties for a given total cost, one can expect that this solution will again look very different from the two previous ones. Therefore one can conclude that in Ando et al.'s (1998) study no aggregation procedure can be thought of that calculates the cost-effective solution from the pure 'ecological' and 'economic' solutions.

The second argument for integrated models, the existence of feedback loops, is even more obvious than the first one. As outlined in Section 3.5, a feedback loop arises if some system component A affects one or more other system components and these feed back into A. Feedback loops can be positive (self-enhancing or destabilising) so that deviations from the equilibrium state are amplified (e.g. Fig. 2.2e,f) or negative (diminishing or stabilising) so that deviations diminish and the system converges to some equilibrium or steady state (e.g. Fig.2.2a,b).

That conservation management can turn a positive feedback loop into a negative one has been distinctly demonstrated by Settle et al. (2002). The authors investigated, with a non-spatial deterministic dynamic model, the interactions between two fish species and human actors in Yellowstone Lake in the United States. One of the two fish species, the cutthroat trout, is native to the lake and popular with hobby fishermen. In the recent past, another species, the lake trout, has invaded the lake and feeds, among other prey, on the cutthroat trout. Over time it has diminished the cutthroat trout population considerably.

Since tourism and fishing for cutthroat trout are relevant sources of income for Yellowstone National Park, the park managers have attempted to control and reduce the lake trout population. Settle et al. (2002) considered three management options: *worst-case*, that is, do nothing and let the lake trout population further increase; *best-case*, that is, fully eradicate the lake trout from the lake; and *policy*, that is, control the lake trout population at an acceptable size.

Next to the managers' interventions, the main interactions considered between the model entities are (Fig. 15.2): the lake trout feeds on the

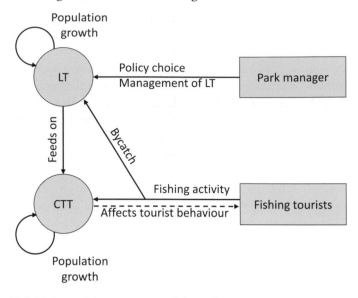

Figure 15.2 Main model components of the Yellowstone Lake model.

cutthroat trout; fishermen who come to the lake are interested only in the cutthroat trout and if this declines they stay away; if by accident a fisherman catches a lake trout, this is taken out of the population.

A feedback loop arises because the fishing effort affects the size of the cutthroat trout population, which in turn affects the fishing effort. This feedback is of a negative, stabilising nature, because a large fishing effort leads to a small population size which reduces the fishing effort, and the population can recover. In the model of Settle et al. (2002) this type of feedback is observed in the *best-case* and the *policy* options, and the analysis of the model shows that the consideration of this feedback in the model leads to more cutthroat trout compared with the case in which this feedback is ignored.

In contrast, in the *worst-case* option, the feedback turns out to be positive and destabilising. The reason for this is that with the decline in fishing effort induced by a decline in the number of cutthroat trout, the accidental by-catch of lake trout is also reduced. As a consequence, the lake trout population can further increase and diminish the cutthroat trout population, so fishing effort declines even more, until no cutthroat trout are left in the lake and fishing effort is zero.

Why does the *worst-case* option lead to such a different outcome than the other two options? In the *best-case* and *policy* options the lake trout is already kept at sufficiently low numbers and the by-catch associated with cutthroat trout fishing is not relevant. Thus, if fishermen stay away due to low cutthroat trout numbers this does not significantly benefit the lake trout, and only the negative feedback is active. In contrast, in the *worst-case* option the lake trout population is rather high and controlled largely through the by-catch. If this by-catch declines due to declining fishing effort the lake trout population increases even more, leading to even fewer cutthroat trout, and generating the positive feedback that eventually leads to the extinction of the cutthroat trout in the lake.

15.2 Challenges: Differing Modelling Approaches

Section 15.1, as well as Section 14.4, provided strong arguments for ecological–economic modelling. However, there are some obstacles to such integrated modelling. As discussed in Part I, models are (or at least should be) designed for given purposes, based on the conceptions of their respective scientific discipline, and may or may not contain various features such as spatial structure, feedback loops and others. One can expect that both purpose and conception have an influence on the structure of a model and the features it contains. Differences between the disciplines' conceptions therefore may imply conceptual and structural differences between the models, which may impede the cooperation of scientists of different disciplines in the development of integrated models (Wätzold et al. 2006).

To analyse the relevance of this issue, Drechsler et al. (2007) carried out a literature review to compare ecological, economic and ecological–economic models. The authors searched eight ecological and environmental-economic journals as well as the interdisciplinary journal *Ecological Economics* for papers between 1998 and 2003 that contain the keywords 'model' or 'model(l)ing' either in title, keywords or abstract, and randomly selected 20 'predominantly ecological', 20 'predominantly economic' and 20 'ecological-economic' papers, where papers were classified in the latter set if they contained at least one reference to an article from an ecological journal such as *Conservation Biology* and one reference to an article from an economic journal such as *Environmental and Resource Economics*.

The 20 models from each set were reviewed with respect to the following questions:

1. Is the model of a general type?
2. Is it specific, that is, applied to a particular case?

3. Is it formulated analytically through mathematical equations or algorithmically through rules?
4. Is it solved/analysed analytically, numerically or via computer simulation?
5. How complex is it (measured by the number of model parameters)?
6. Is it non-spatial, spatially differentiated, spatially explicit or differentiated and explicit (for the definitions, see Section 3.1)?
7. Is it static, dynamic with continuous time or dynamic with discrete time?
8. Does it consider uncertainty (either in the form of stochasticity or via sensitivity analyses or both)?

Table 15.1 shows how many of the reviewed ecological, economic and ecological-economic models contain which characteristic. From the resulting counts and some additional correlation analyses, Drechsler et al. (2007) identified two main groups of models into which all of

Table 15.1 *Characteristics of the reviewed ecological, economic and ecological-economic models, taken from Drechsler et al. (2007). All numbers except for the means and standard deviations of the numbers of model parameters represent percentages rounded to multiples of 5. Note: some articles included a general and a specific model analysis, so the summed percentages of general and specific models exceed 100.*

Characteristic	Ecological	Economic	Ecological-economic
General	40	90	90
Specific	85	25	80
Equation-based	75	100	90
Rule-based	25	0	10
Analytical solution	20	100	55
Numerical solution	30	0	20
Simulation	50	0	25
Mean no. of parameters	16.5	8.8	11.9
Std. dev. of no. of parameters	9.9	5.4	9
Non-spatial	55	75	60
Spatially differentiated	10	25	15
Spatially explicit	0	0	15
Differentiated and explicit	35	0	10
Static	30	50	15
Continuous time	5	10	20
Discrete time	65	40	65
Uncertainty included	97	50	66

the 60 models can be sorted: 'simple models' that are general, formulated through equations, analysed analytically, have few parameters, are static or continuous in time, are often deterministic and do not consider spatial structure; and 'complex models' that are usually specific, more frequently rule-based and analysed via simulation, have many parameters, are often dynamic but with discrete time and consider spatial structure. Almost all economic models were of the first type while the second type was found most often in ecological models. The ecological-economic models fell in between the 'simple' and 'complex' models.

Since the review of Drechsler et al. (2007) may now be outdated, I carried out a similar review, based on articles from the three journals *Conservation Biology* (CB), *Environmental and Resource Economics* (ERE) and *Ecological Economics* (EE), from the years 2013–17. From each journal I randomly selected 20 articles:

- CB: Aiello-Lammens and Akçakaya (2016), Baker et al. (2016), Baskett and Waples (2012), Bull et al. (2014), Di Minin et al. (2014), Doak et al. (2013), Drechsler et al. (2016), Gjertsen et al. (2014), Huang et al. (2017), Iacona et al. (2016), Lentini et al. (2013), Ng et al. (2014), Nuñez et al. (2013), Rout et al. (2017), Ryberg et al. (2014), Simmons et al. (2015), Spencer et al. (2017), Tenan et al. (2016), Visconti and Joppa (2014), Wilson and Hopkins (2013);
- ERE: Annachiarico and Di Dio (2017), Asheim (2013), Atewamba and Nkuiya (2017), Chavas and Di Falco (2017), Creti et al. (2018), Egger and Nigai (2015), Gahvari (2014), Gerigk et al. (2015), Hatcher and Nøstbakken (2015), Jiang and Koo (2014), Kakeu and Johnson (2018), Kalkuhl et al. (2015), Lehmann and Söderholm (2018), Pautrel (2015), Raffin and Seegmuller (2017), Rämö and Tahvonen (2017), Rivers and Growes (2013), Stoeven (2014), Tahvonen et al. (2013), Thompson (2013);
- EE: Bamière et al. (2013), Bauer and Wing (2016), Comello et al. (2014), Dalmazzone and Giaccaria (2014), Drechsler (2017c), Eisenack (2016), Ge et al. (2015), Grovermann et al. (2017), Hackett and Moxnes (2015), Iho et al. (2017), Jalas and Juntunen (2015), Kragt and Robertson (2014), Pezzey and Burke (2014), Piñero et al. (2015), Reznik et al. (2017), Roman et al. (2017), Schöttker et al. (2016), Soltani et al. (2014), Thomas and Azevedo (2013), Zhu et al. (2015),

and evaluated them in the same manner as Drechsler et al. (2007). However, the analysis differs slightly from the previous one in that the statistics are now based on journal affiliation and not the classification as one of the three model types 'ecological', 'economic' and 'ecological-economic'.

To allow for comparability with Drechsler et al. (2007) I further classified the 60 models as one of these three types but used a stricter criterion for 'ecological-economic'. Models with a focus on ecology, including most of those from *Conservation Biology*, were classified as ecological-economic only if they consider economic issues in a significant way, such as economic costs or the execution of a cost-effectiveness analysis. Economic models, including most of those from *Environmental and Resource Economics* and a large portion of those from *Ecological Economics*, were classified as ecological-economic only if an environmental resource is explicitly modelled as an (evolving) stock, such as CO_2 concentration in the atmosphere, level of water pollution or population size. In contrast, models that for instance only contain a cost function that models the abatement cost for some pollutant were classified as 'economic'. Another difference with the previous review is that I distinguished between the consideration of stochasticity in the model dynamics (cf. Chapter 5) and uncertainty in the sense that uncertain model parameters are varied and associated changes in model outputs are recorded (sensitivity analysis). The results for the first and second steps of the analysis are given in Tables 15.2 and 15.3.

The new review, Table 15.3, largely confirms the findings of the previous one by Drechsler et al. (2007). Economic models are to a higher proportion general, equation-based, analytically solved, non-spatial, static and deterministic, while the ecological models are largely specific, solved numerically or through simulation, consider spatial structure, are more often dynamic and are usually stochastic. Moreover, similar to the previous review, even most of the ecological models are equation-based and the analysis of all three types of models usually addresses uncertainty.

However, there are also a few differences between the two reviews. Firstly, the model analyses of the new review are less often general, which might be due to a stricter interpretation of generality. Most of the papers investigated do contain a general model formulation (e.g. through equations) which then may be followed by a parameterisation of the model with specific parameter values. Despite their general formulation I still classified these models as specific because they were evaluated only for a specific parameterisation; meanwhile, I classified only those model analyses as general that included, for example, sensitivity analyses beyond a specific parameterisation to generate general insights.

Table 15.2 *Characteristics of 60 (20 per journal) randomly sampled models from* Conservation Biology *(CB),* Environmental and Resource Economics *(ERE) and* Ecological Economics *(EE). All numbers except for the means and standard deviations of the numbers of model parameters represent percentages rounded to multiples of 5. Note: The count of model parameters is highly subjective. In terms of parameter number, multi-stage population models and multi-species models were counted as two-stage and two-species models, respectively; models with more than two land-use types were counted as models with two land-use types; and the spatial structure of habitat networks was represented by three parameters: sizes of two habitat patches and the distance between the two patches. These choices were made to avoid an 'inflation' of parameter numbers that does not necessarily reflect an increase in model complexity.*

Characteristic	CB	ERE	EE
Ecological–economic	35	60	80
General	25	75	25
Specific	95	50	75
Equation-based	80	100	95
Rule-based	20	0	10
Analytical solution	0	75	10
Numerical solution	60	35	60
Simulation	40	15	40
Mean no. of parameters	10.1	10.9	16.9
Std. dev. of no. of parameters	5.2	5.7	9.1
Non-spatial	55	85	70
Spatially differentiated	15	15	15
Spatially explicit	0	0	0
Differentiated and explicit	30	0	15
Static	30	30	45
Continuous time	5	50	25
Discrete time	65	20	30
Stochasticity	60	15	20
Uncertainty considered	70	65	60

Secondly, in all three model types the proportion of analytical analyses is smaller than in Drechsler et al. (2007) and numerical solutions and simulations were applied more often. Thirdly, the number of parameters in the ecological models in the new review is smaller than in the previous one, while in the ecological-economic models it is larger. The former observation may be due to a different counting of the parameters (which is very subjective and ambiguous), so that in the new review, for instance, a stage–based ecological model with more than two stages was counted to have two stages because two stages are likely to produce more complex dynamics than a single stage but the inclusion of more than two stages may not add substantially more complexity. The second observation might be of an objective nature such that developers of ecological–economic models attempt to maintain the full complexity in both subsystems, so the integrated model is more complex altogether. Lastly, the proportion of static samples in the economic and ecological–economic models has somewhat increased between the two reviews.

Very similar results are obtained when considering the three journals rather than the three model types (Table 15.2). The main differences include: (i) the proportion of general models in ERE is larger than in the class of economic models and the proportion of general models in EE is smaller than in the class of ecological–economic models, which seems to be due to the smaller number of general models in EE. This also may explain that (ii) some economic models are rule-based, but none of those are from ERE. (iii) A larger proportion of models from CB is spatially

Table 15.3 *Characteristics of the ecological (*n = 13*), economic (*n = 12*) and ecological-economic (*n = 35*) models found among the 60 sampled models of Table 15.2. Details as in Table 15.2.*

Characteristic	Ecological	Economic	Ecological-economic
General	15	60	45
Specific	100	60	70
Equation–based	85	100	90
Rule-based	15	10	10
Analytical solution	0	75	25
Numerical solution	60	40	50
Simulation	40	10	40
Mean no. of parameters	9.0	10.4	15.6
Std. dev. of no. of parameters	3.9	5.7	9.6
Non-spatial	55	80	70
Spatially differentiated	5	10	20
Spatially explicit	0	0	0
Differentiated and explicit	40	10	10
Static	25	70	30
Continuous time	5	15	35
Discrete time	70	15	35
Stochasticity	60	10	30
Uncertainty considered	85	75	55

differentiated compared with the set of ecological models, which is obviously due to the ecological-economic models found in CB. (iv) Some economic models are spatially differentiated and explicit – in contrast to the models from ERE, which is due to the economic models found in EE. (v) The proportion of static models in the set of economic models is larger than that in the models from ERE, and the proportion of static models in the set of ecological-economic models is smaller than that in the models from EE, which seems to be due to some static models found in EE. A final and significant observation is that the share of ecological-economic models is highest in EE (80 per cent), followed by ERE (60 per cent), and lowest in CB (35 per cent).

To conclude, as in the review of Drechsler et al. (2007) economic models are (still) to a much higher proportion non-spatial, static and deterministic than ecological models. This seems to reflect the continuing prevalence of the paradigm of (market) equilibrium in economics, including environmental economics. The relatively large share of analytical model analyses may be due to the fact that equilibria and steady states are more easily calculated and analysed than dynamic (and spatial) patterns, as well as the author's (admittedly subjective) impression that economic curricula at universities generally involve more training in mathematical analysis and algebra than do ecological ones.

If one recalls the requirements for integrated modelling formulated in Section 14.4, this dichotomy between ecological and economic models, and in particular the widespread neglect of spatial structure, dynamics and stochasticity in economic models, is quite problematic. At the same time, the ecological-economic models demonstrate that it is possible to bridge the gap between ecological and economic models and, for example, integrate spatio-temporal dynamics and stochasticity in the analysis of market-based conservation instruments (see more in Chapters 16 and 17). This integration seems to come at the cost of increased model complexity, but this cost might be relativised by improvements in the analysis of complex models such as pattern-oriented modelling (Grimm et al. 2005; Hartig et al. 2011), sensitivity and robustness analysis (Ben-Haim 2001; Saltelli et al. 2009) and multivariate analysis (Pielou 1984; Johnson and Wichern 2013), which can all be carried out with increasing efficiency as computer power continues to grow.

15.3 Concepts That Facilitate Integration

Although ecological and economic concepts and models differ, there is also considerable overlap between the two. Proops (1989) lists a number of examples, some of which go back centuries. They include (for references, see Proops (1989)) analogies between human and insect societies, the use of game theory to explain behaviour of humans and animals in social and ecological science, respectively, the mapping and modelling of flows in ecological and economic systems, the use of market concepts to describe ecosystem dynamics, and the concept of evolution that can be found in evolutionary biology as well as evolutionary economics.

Further similarities concern the notion of resource scarcity (Baumgärtner et al. 2006). In addition, Optimal Foraging Theory (e.g., Begon et al. 1990, ch. 9.11) deals with optimal behaviour in the search for and use of scarce resources by biological organisms. Much of the terminology used in that theory can also be found in textbooks on micro- or behavioural economics. Although economics tends to focus more on steady states and equilibria than does ecology, which also addresses oscillations, stochastic fluctuations and other more complex dynamic patterns, stability in its broadest sense is an important concept and topic of research in both disciplines. The main difference seems to be that ecology knows far more stability concepts than does (neoclassical) economics, such as persistence (cf. Chapter 5), resilience (cf. Section 14.3) and others (Grimm 1997).

Lastly, a factor that cannot be overestimated is the integrative ability of mathematics. Many ecological and economic concepts are formulated formally, either through equations, rules or graphs (which can usually be cast into mathematical rules or equations), and most of the integrating concepts mentioned previously are formulated in the language of mathematics. Even though ecological and economic terminologies generally differ, the common language of mathematics can bridge these differences. Since mathematics is (almost) free of contradictions, the most difficult part in the integration of ecology and economics through mathematics is the correct translation of the respective concepts into mathematics. Once this has been accomplished, the rest of the modelling exercise is the 'mere' application of mathematical methods and tools.

16 · Integration of Ecological and Economic Models

16.1 A Classification of Integration in Ecological-Economic Models

The present chapter deals with the integration of ecological and economic modelling, focusing on the influences between the ecological and economic subsystems. Influences *within* the two subsystems are topics of the disciplinary ecological and economic models and are not considered in this chapter. The main characteristic of ecological-economic model integration is whether each subsystem affects the respective other subsystem ('bidirectionality') or whether only one subsystem affects the other without an influence in the reverse direction ('monodirectionality').

Both mono- and bidirectional models can be analysed according to four main concepts:

1. cost-effectiveness and budget-effectiveness (cf. Section 11.3), which states that an ecological benefit is maximised for a given total cost or conservation budget, respectively (or the total cost or the budget to reach a given level of ecological benefit is minimised);
2. efficiency (cf. Section 9.1), which adds to the cost- and budget-effectiveness concepts the question of the welfare-maximising level of cost or budget and the associated welfare-maximising level of the ecological benefit;
3. resilience (cf. Section 14.3), which measures the ability and conditions under which a system maintains its characteristic structure and behaviour; and
4. sustainability (cf. Section 14.3), which adds to the resilience concept the judgement that some system structures and behaviours are preferable to others.

The following sections address these characteristics of model structure and analysis. Section 16.2 presents monodirectional models with influence of the economic on the ecological subsystem, analysed according to the concept of cost-effectiveness; Section 16.3 presents bidirectional models analysed with a focus on resilience and sustainability; Sections 16.4–16.6 (briefly) introduce economic valuation to move from cost-effectiveness to efficiency; and Section 16.7 reflects some of these concepts in the context of natural capital.

16.2 Economy Affects Ecology

This section presents examples of integrated ecological–economic models in which economy and land use affect ecological outcomes but not vice versa (examples of ecological–economic models that consider the influence of the ecological subsystem on the economic subsystem without considering an influence in the opposite direction are not known by the author). As may be typical for this kind of model, they consider the cost-effective allocation of conservation resources, which may be via planning or via a market-based instrument (cf. Chapter 9). The first example is a rather old but influential and frequently cited model study by Wu and Boggess (1999), who analyse the cost-effective allocation of a conservation budget C_{tot} in two regions ($i = 1,2$) to maximise some total ecological benefit B_{tot}. The benefit B_i in each region is modelled as a function of the conserved resource, for example, the amount of protected area A_i, which in turn is a function of the budget C_i spent in region i. For additive benefits, $B_{tot} = B_1 + B_2$, with concave benefit functions $B_i(A_i)$ and convex cost functions $C_i(A_i)$, this resource allocation problem has been presented in Section 9.2.

Wu and Boggess (1999) add two important features to the ecological benefit function B_i. Firstly, B_i need not be concave in A_i (with the marginal benefit dB_i/dA_i monotonically declining with increasing A_i), but it may be sigmoid, that is, convex for small areas A_i and concave for larger A_i. This relationship represents ecological threshold behaviour where a conservation effort yields a measurable ecological benefit only if it exceeds a certain minimum value. The second issue addressed by Wu and Boggess (1999) is spatial interactions or externalities by assuming that each of the two ecological benefits B_1 and B_2 poses a positive externality on the benefit in the respective other region, measured by a function αB_i^{δ}. As the authors demonstrate, these two additional features change the simple allocation rule derived in Section 9.2 substantially.

Next to Ando et al. (1998), which represents a second example of cost-effective conservation resource allocation, a third seminal paper in this field is by Polasky et al. (2008). The authors optimise the spatial allocation of 14 land-use measures in the Willamette Basin, Oregon, such as pasture, cropping agriculture, forestry, rural residential use, as well as several conservation measures applied to natural habitats in the region such as prairie, old-growth conifer and marsh.

Each land parcel in the study region can be assigned one of the 14 land-use measures. Economic revenues are determined for each land-use measure and each land parcel, taking into account, for instance, agricultural productivity and exogenous market prices for agricultural products (agriculture), age of timber stand and exogenous timber price (forestry), proximity to urban areas and existing building density (residential use) and costs of carrying out conservation measures (conservational use). Adding up the economic revenues and subtracting the costs yields a total societal income for the chosen spatial allocation of land-use measures in the study region.

Each land-use measure in each land parcel is assigned a suitability value for each of the 267 vertebrate species to be considered. Adjacent suitable land parcels are aggregated to habitat patches whose carrying capacities are calculated. Species-specific spatial connectivity of these habitat patches is evaluated, taking the pairwise distances between the patches and the dispersal ranges of the species into account. From the carrying capacities of the habitat patches and the connectivity index, an expected number of breeding pairs in the study region is calculated for each species and valued through a concave ecological benefit function. Adding up these benefit functions delivers a total ecological benefit for the chosen spatial allocation of land-use measures.

With these ecological and economic models each land-use pattern, defined by a particular spatial allocation of the 14 land-use measures over the 8,176 land parcels in the study region, delivers a number of conserved species and a societal income. Plotting the number of species versus societal income leads to a graph as in Fig. 16.1. Each dot in the graph represents a land-use pattern.

Some of the dots (e.g. the point (0.26, 1.50)) represent inefficient land-use patterns in the sense that a higher ecological output (number of species in Polasky et al (2008)) could be achieved at the same or even higher economic output (societal income in Polasky et al (2008)), or vice versa, a higher economic output could be achieved without reducing the ecological output. Only the points on the surface of the point cloud represent land-use patterns that maximise the ecological output for given economic output. The set of these points is termed the production possibility frontier (PPF).

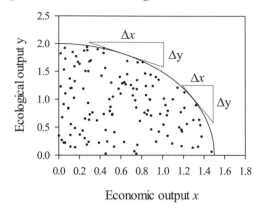

Figure 16.1 Fictitious combinations of some economic (e.g. societal income, x) and ecological (e.g. number of conserved species, y) output drawn randomly under the constraint $(2x/3)^2 + (y/2)^2 \leq 1$.

The downward sloping shape of the PPF indicates a trade-off between the two criteria in the sense that to increase one of them the other must be reduced. The slope at a given point on the PPF is termed the marginal rate of transformation, because it indicates how much ecological output must be sacrificed to increase the economic output by one unit, or vice versa, how much economic output must be sacrificed to increase the ecological output by one unit. One can see that at a currently high ecological output and correspondingly low economic output (upper left end of the PPF) a large gain in economic output can be achieved with a rather small reduction of the ecological output, or conversely, a small increase in the ecological output incurs a substantial reduction in the economic output. The opposite is found for the lower right end of the PPF where economic output is high and ecological output is low.

Considering that a decrease in the economic output (societal income) can be regarded as an increase in economic (opportunity) costs, it is equivalent to the concept of cost-effectiveness that demands the ecological output (number of conserved species) to be maximised under a cost constraint. If that cost constraint is varied and the associated cost-effective ecological output is plotted against the cost constraint, a cost function is obtained that is equivalent to the PPF (cf. Section 16.6).

An extension of Polasky et al. (2008) is the paper by Nelson et al. (2008), who relax the assumption of a perfectly informed conservation planner and instead induce the land-use pattern through conservation payments (cf. Section 9.3). This adds another level of complexity, because the model considers not only the influence of the land-use pattern on the ecological system but also the influence of the payment scheme, via the decisions of the landowners, on the land-use pattern. To model the decisions of the landowners, an econometric (statistical) model was fitted to the behaviour of landowners observed in the real world (Lewis et al. 2011).

16.3 Models with Bidirectional Influence

A 'bidirectional' model in which the ecological subsystem affects the economic subsystem and vice versa via a feedback loop is considered in the model analysis by Settle et al. (2002) described in Section 15.1. The size of the cutthroat trout population affects, via preference functions of the tourists, the number of tourists fishing in Yellowstone Lake. Since the number of fishing tourists affects the size of the cutthroat trout population, a feedback loop exists between the ecological and economic subsystems that couples the dynamics in the fish population sizes and the number of tourists.

The consideration of feedback loops requires broadening the view from cost-effectiveness analysis to an analysis of resilience and sustainability. In a way, Settle et al. (2002) analyse the resilience of the system of tourists and cutthroat trout to the invasion of the lake trout and find that if the lake trout become too abundant, the stabilising negative feedback loop between the number of cutthroat trout and the number of fishing tourists turns into a positive and destabilising one, which eventually leads to the extinction of the cutthroat trout in the lake. Since the cutthroat trout is endangered in Yellowstone Lake and the revenues from fishing are an essential source of income for Yellowstone National Park, the management problem being analysed is also one of sustainability.

Another instructive example of a feedback loop that is formulated entirely through mathematical equations is the model by Bulte and Horan (2003). The authors analyse a resource management problem in sub-Saharan Africa in which a local community can use its land L for growing crops on an area A and conserving wildlife on an area H with area constraint

$$A + H = L. \tag{16.1}$$

To use their land, people spend their total (working) time T managing their crops (W) and hunting (E):

$$W + E = T. \tag{16.2}$$

The question analysed with the model is what steady state the system will assume and, in particular, whether that steady state will be dominated by agriculture or by wildlife conservation.

To start with the economic subsystem, the community draws income from wildlife by hunting and selling products from the hunted animals at an exogenous price p per animal. According to the Schaefer equation (Hanley et al. 2007, p. 271), which is frequently used in the modelling of renewable resource management, the number of animals hunted, h, is given by the product of wildlife population size x, hunting effort E and a catchability coefficient q that measures how easily the animals can be caught:

$$h = xqE. \tag{16.3}$$

With ph being the revenue from selling the animal products and E being the cost of supplying these products, the return from wildlife π_H, that is, the ratio of revenue and cost, equals

$$\pi_H = \frac{pqxE}{E} = pqx. \tag{16.4}$$

In addition to hunting the community grows crops, the amount of which depends on the agricultural land A in a quadratic manner (for the economic explanations, see Bulte and Horan (2003))

$$C = cA^2, \tag{16.5}$$

while the time spent for cropping increases linearly with area

$$W = \alpha A. \tag{16.6}$$

Assuming that the crops are sold on thin local markets at an endogenous price that decreases with an increasing amount of crops according to $C^{1-\beta/2}$ with parameter $\beta \in (0,1)$, the revenue from growing crops is given by $\alpha^{\beta/2} A^\beta$ and the return is

$$\pi_A = \frac{\alpha^{\beta/2} A^\beta}{W} = \alpha^{\beta/2-1} A^{\beta-1}. \tag{16.7}$$

The ecological subsystem is modelled through a logistic growth equation of the wildlife population, with intrinsic growth rate γ and carrying capacity H (cf. Section 4.2):

$$\frac{dx}{dt} = \gamma x(H - x) - h, \tag{16.8}$$

where the growth is reduced by the amount h of animals harvested. Using the constraints of Eqs (16.1) and (16.2), habitat area H and effort E can, with Eqs (16.3) and (16.6), be replaced to obtain

$$\frac{dx}{dt} = \gamma x(L - A - x) - xq(T - \alpha A) \tag{16.9}$$

which describes the impact of the economic subsystem, in particular the choice of agricultural area A, on the dynamics of the ecological subsystem represented by the wildlife population x. The feedback from the ecological to the economic subsystem is established by assuming that the community strives for a maximum return and expands the agricultural area if the return π_A from agriculture exceeds the return from wildlife π_H and reduces it otherwise:

$$\frac{dA}{dt} = \eta(\pi_A - \pi_H), \tag{16.10}$$

where parameter η measures how fast the community responds to a given difference between the returns π_A and π_H. The feedback loop is now complete: the returns π_A and π_H depend on the size of the wildlife population, x (and on agricultural area A), and determine the change in agricultural area A. Via Eq. (16.9) agricultural area in turn affects the change in the wildlife population x.

The authors determine the steady state of the system by setting the time derivatives in Eqs (16.9) and (16.10) to zero and solving for the levels of x and A. For plausible values of the model parameters, three steady states are identified: one with maximum agricultural area but without any wildlife, one with a large amount of wildlife and little agricultural area, and one with both state variables in between.

A stability analysis (for an outline, see, e.g., Hanley et al. (2007, ch. 9.8)) that analyses how the system develops from an arbitrary state (x, A) which differs from the three mentioned steady states reveals that the steady state with intermediate levels of wildlife and agricultural area is unstable, so that in the course of time x and/or A move further away from their steady state values. Depending on whether the current amount of wildlife is larger or smaller than its level in the intermediate steady state (fig. 1 in Bulte and Horan (2003)), the system develops into the steady state with much wildlife and little agricultural area or into the steady state with much agricultural area and no wildlife. Again, these results contribute to discussions about resilience and sustainability, since they demonstrate that, depending on its initial state, a system can move into very different states, and that some of these states may be unfavourable.

Both feedback loops outlined here arise because biodiversity (the cutthroat trout and the wildlife population) have a value to the agents (the fishing tourists and the local community). If biodiversity has no value, self-interested agents, as they are assumed in neoclassical economics, will not respond to changes in biodiversity, leading to the market failure described in Section 9.1. As discussed, the market failure can be fixed (to some extent) by market-based instruments that create an artificial demand for biodiversity (Sections 9.3 and 9.4). The demand function assigns biodiversity a price, so landowners can draw economic benefits from providing it. If that price is assigned to conservation *efforts*, such as conserved area as in the examples of Section 16.2, there will still be no feedback from the ecological to the economic subsystem, because the landowners respond only to the conservation policy and not to changes in the biodiversity. If, in contrast, the price is on *biodiversity* itself, the ecological subsystem affects the economic subsystem and a feedback loop can emerge. This is the case in so-called output-based payments under which a landowner receives a payment if a particular ecological outcome, such as the presence of a target species on their land, is observed. The dynamics of an output-based payment will be analysed in Section 17.3.

If biodiversity conservation is driven by policy instruments, another feedback loop may arise beyond that between the economic subsystem of the landowners and the ecosystem, in this case between these subsystems and the subsystem formed by the policymakers (Polasky and Segerson 2009). Such a feedback arises, for example, if the success or failure of a policy instrument triggers changes in that policy instrument to improve its performance.

16.4 Social Efficiency of Biodiversity Conservation via Indifference Curves

The production possibility frontier (PPF; cf. Fig. 16.1) is an integrative concept by definition, since it relates the maximum ecological and economic outputs that can be achieved simultaneously. As explained, its slope represents the marginal rate of transformation that tells how much of one criterion, for example economic output, must be sacrificed to increase the other criterion (ecological output). From the point of view of efficiency or social welfare, however, the PPF only represents one side of the coin. The other side is the question of how much economic output society is *willing* to sacrifice to increase the ecological output by a certain amount (in Section 16.6 we will see that this question is equivalent to the question of which level of ecological output equates marginal demand and marginal supply).

For the case of an individual decision-maker, this so-called marginal rate of substitution (between two criteria) has been introduced in Fig. 12.1 in Section 12.1. There it is represented by the slope of the iso-utility line that includes all combinations of quantities x and y that lead to the same utility, or well-being, in the decision-maker. The iso-utility curve is often termed an indifference curve, since it includes all combinations of x and y, to which the decision-maker is indifferent, so they do not prefer any particular point on the curve to any other. Clearly, indifference curves further away from the origin, containing points with larger x and/or larger y, are associated with higher utility and are preferred to indifference curves closer to the origin (assuming here, of course, that the criteria x and y are to be maximised).

How can the indifference curve and the information about the marginal rate of substitution be used to support decisions, and in particular determine the efficient, welfare-maximising levels of ecological and economic output? To illustrate the procedure, Fig. 16.2a shows the PPF of Fig. 16.1,

$$\left(\frac{2x}{3}\right)^2 + \left(\frac{y}{2}\right)^2 = 1 \qquad (16.11)$$

(stating that certain constraints limit economic (x) and ecological (y) outputs), together with the indifference curve of Fig. 12.1a

$$U = \alpha x + \beta y \qquad (16.12)$$

with $\alpha = 2$ and $\beta = 1$ for various levels of utility U_0.

A decision-maker, be it a self-interested individual or a representative of society, aims at maximising utility U. In Fig. 16.2a, utility levels above $U_{\text{eff}} = 13^{0.5} \approx 3.61$ (dotted lines) would violate the production constraint, Eq. (16.11), and are not feasible. Utility levels below U_{eff} (dashed lines) are feasible but inefficient, since higher utilities could be generated without violating the production constraint. The maximum feasible, efficient level of utility U_{eff} (solid line) is generated by the unique solution $x^* = 9/2/13^{0.5} \approx 1.25$ and $y^* = 4/13^{0.5} \approx 1.11$, represented by the point where PPF and indifference curve are tangent to each other. The utility-maximising levels x^* and y^* are the efficient levels of

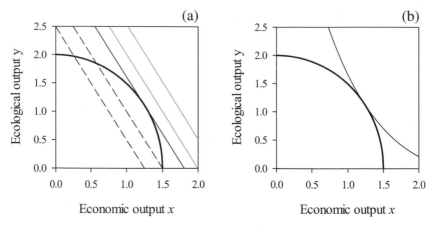

Figure 16.2 Production possibility frontier, $(2x/3)^2 + (y/2)^2 = 1$ (bold lines), and indifference curves $U(x,y) = U_0$ (thin lines). Panel (a): $U = 2x + y = U_0$ with $U_0 = 2.5, 3.0, 3.61, 4.0, 4.5$ (from left to right); panel (b): $U = 2x^{0.5} + y^{0.5} = 3.29$.

ecological and economic output, respectively. Analytically, they are obtained in a straightforward manner by maximising $U = 2x + y$ (Eq. 16.12) under the constraint of Eq. (16.11).

This result can be cast into a general rule of efficiency. Consider that the tangency of two curves implies that the slopes of the two curves at the tangent point are identical. The slope of the PPF was explained to be the marginal rate of transformation, and the slope of the indifference curve is the marginal rate of substitution. Therefore the point of efficiency is defined by equality of the marginal rate of transformation and the marginal rate of substitution (Varian 2010).

This sounds very similar to the efficiency rule deduced in Section 9.1 that the efficient level of biodiversity is where the marginal cost (the slope of the economic cost function) equals the marginal benefit (the slope of the ecological benefit function). In Section 16.6 I will show that these two conditions for efficiency are indeed mathematically equivalent.

For a given PPF, the marginal rate of substitution, that is, the slope of the indifference curve, strongly affects the efficient levels of x and y. In the case represented by Fig. 16.2a the slope of the indifference curve is rather steep, leading to a rather large x^* and rather small y^*. This agrees with the parameterisation of the utility function that gives x a higher 'weight' ($\alpha = 2$) than y ($\beta = 1$), implying that y would have to be increased by two units to compensate for a decrease of x by one unit. A smaller x^* and a larger y^* would be obtained if the indifference curve had a shallow slope. In the previous considerations the indifference curve was assumed to be a straight line. The arguments, however, apply equally to non-linear indifference curves such as the one in Fig. 16.2b.

Analyses of this type can be regarded as what is termed multi-criteria analysis (MCA) (e.g., Ishizaka and Nemery 2013). MCA is used to rank a (finite or infinite) set of decision alternatives with regard to a number of objectives, each measured by a criterion. The previous analysis considers a rather simple case of two criteria, ecological output and economic output, which may represent the two policy objectives of economic prosperity and environmental quality. The critical step in MCA is the aggregation of the criteria. Most methods of MCA attach weights to the criteria and build some aggregate score from the weighted criteria. The most common method of MCA is Multi-Attribute Value Theory (MAVT), which multiplies the performances of the alternatives with the weights and adds the products obtained. For the case of two criteria and with the performances of the alternatives in these criteria denoted as x and y, Eq. (16.12) represents an MAVT with criteria weights α and β.

Sometimes the performances of the alternatives are mapped into so-called partial utilities before multiplication with the weights and summation into a total utility. An example is Eq. (12.2), which is the basis of the indifference function in Fig. 16.2b. This method is called Multi-Attribute Utility Theory (MAUT) (Ishizaka and Nemery 2013, pp. 81ff.).

16.5 Social Efficiency of Biodiversity Conservation via Demand Functions

The second main approach for the identification of the efficient level of conservation effort and associated utility is to establish a benefit and a cost function as in Fig. 9.1 and maximise the difference between benefit and cost. If benefit and cost are, for example, expressed as functions of conserved area as in Fig. 9.1, the maximisation delivers the efficient level of conserved area and the associated ecological benefit and economic cost (Eqs (9.1) and (9.2)).

In order to make Eq. (9.1) mathematically meaningful, benefit and cost must, of course, be measured on the same dimension, which is often monetary units such as euros. This requires the ecological benefit, such as species richness, presence of a target species and so forth, to be monetised and translated into monetary units.

Monetary, or economic, valuation of environmental goods and services is a major research field in environmental economics (Edwards-Jones et al. 2000, chs 8–11; Hanley et al. 2007, ch. 11). The two main approaches are direct and indirect valuation. In indirect valuation methods, the value of the environment is inferred from observable decisions, including market decisions, of humans. For instance, the amount people spend to travel to a nature reserve can be used to estimate the value of that reserve to these people ('travel cost method'), or the statistical influence of the presence of a nearby nature park on house prices can be used to estimate the value of that park.

In direct valuation methods, respondents are confronted with a hypothetical decision situation and asked how much money they would be willing to spend for an improvement of some environmental good such as biodiversity or air quality ('willingness to pay', WTP), or how much monetary compensation they would demand for the reduction of that good ('willingness to accept', WTA). Assessing WTP and WTA is a delicate task and requires a great deal of care in the set-up of the valuation procedure, such as the appropriate provision of the necessary background information to the respondents, the adequate sequence of the questions asked to the respondents and many more (Edwards-Jones et al. 2000, ch. 8; Hanley et al. 2007, ch. 11.2).

The two main approaches to direct valuation are the contingent valuation method (CVM) (e.g. Mitchell and Carson 1989) and choice experiments (CE) (e.g. Louviere et al. 2000). In the CVM, the respondents are usually confronted with a set of prices from which they are asked to select one that best fits their personal WTP and/or WTA. The WTP (or WTA) obtained for the respondents are then evaluated statistically, and the average is multiplied with the number of individuals or households in the region of concern to obtain the regional monetary ecological benefit.

While in the CVM respondents value only a single environmental attribute at a time, in CE they assess combinations of attributes such as {species richness, water quality, number of jobs created} plus a monetary attribute such as increase of some tax. For this the respondents are confronted with several bundles that differ in the levels assumed by the attributes. In the present example, such a bundle could be, for example, {14 species conserved, high water quality, 250 jobs created and tax increased by 10 euros per month}. The respondents are then asked to rank the bundles, and through a statistical analysis the effects of the attributes and their levels on the people's ranking decisions are estimated. Dividing the former effects by the effect of the monetary attribute allows for expressing those former attributes on a monetary scale.

An example of an efficiency analysis in the field of biodiversity conservation is Wätzold et al. (2008). It considers the conservation of an endangered butterfly species, *Maculinea teleius*, in a region in Rhineland-Palatinate, Germany. The species inhabits extensively managed grassland and relies on an adequate mowing regime, defined by the frequency and the timing of cuts, for its survival. A cost function $C(A)$ was established that indicates how costly it is to manage an area of size A in a species-friendly manner. It is comparable to that in Fig. 9.1 (short-dashed line) but of linear shape (i.e. marginal cost is constant). Through CVM, the WTP of the people in the region was assessed for three different conservation projects. The projects differed in their spatial extent, with conserved areas of 4 ha, 16 ha and 64 ha. The estimated regional monetary benefit $B(A)$ and the derived marginal benefit $\partial B/\partial A$ are comparable to those in Fig. 9.1 (solid line).

The authors found that within the range of conserved area considered, $A \leq$ 64 ha, benefit and marginal benefit exceed cost and marginal cost, respectively: $B(A) > C(A)$ and $\partial B(A)/\partial A > \partial C(A)/\partial A$, which implies that the socially efficient level of conserved area is larger than 64 ha. From a conservational point of view this is, on the one hand, a positive result and agrees with earlier findings from other authors that in Germany the observed (marginal) demand for conservation would call for an expansion of conservation efforts. On the other hand, one may question what the respondents were really valuing when selecting a certain WTP for the butterfly. Were they thinking only of *M. teleius*, or rather of all grassland butterfly species, or all grassland species? If in a second interview the respondents were asked for their WTP to conserve a second butterfly species, would they still state the same values, and the same in further interviews for the dozens of other species? Such ambiguities are a major problem in the direct valuation of environmental benefits and the interpretation of the measured WTP and WTA. Nevertheless, even though the WTP and WTA measured in economic valuation studies are subject to uncertainty, they are highly important in political and economic decision-making, because they make the value of the environment explicit and can guide political and economic decision-making.

16.6 Equivalence of Indifference Curves and Demand Functions and Conclusions for Environmental Valuation

In the present section I will show that the assessment of social efficiency via indifference curves (IC) (as in Fig. 16.2) is equivalent to its assessment via demand curves (as in Fig. 9.1). For this I first show that the formulation of a decision problem through benefit and cost functions is mathematically equivalent to the formulation of the problem via a PPF and an IC. In a second step I will show that the optimal solution of the decision problem derived via PPF and IC is identical to the solution derived by maximising the difference between benefit and cost.

To start with the first step, denote by A the level of some environmental good such as biodiversity. For instance, A may be the number of species conserved in a region (Polasky et al. 2008) or area inhabited by some endangered butterfly species (Wätzold et al. 2008). Assume provision of A incurs costs to society of magnitude $C(A)$ with $\partial C/\partial A > 0$ and $\partial^2 C/\partial A^2 > 0$, and generates a monetary benefit $V(A)$ with $\partial V/\partial A > 0$ and $\partial^2 V/\partial A^2 < 0$ (cf. Section 9.1).

For a numerical example, consider

$$C(A) = 0.05A + 0.1A^2 \tag{16.13}$$

and

$$V(A) = 3A^{0.5} \tag{16.14}$$

(cf. Fig. 9.1). In Section 9.1 it was shown that the efficient level of A that maximises social welfare, $W(A) = V(A) - C(A)$, is obtained by the equality of marginal benefit and marginal cost, $\partial V/\partial A = \partial C/\partial A$. In the numerical example of Eqs (16.13) and (16.14) this reads $0.05 + 0.2A^* = 1.5(A^*)^{-0.5}$, implying an efficient biodiversity level of $A^* \approx 3.67$.

Cost C is forgone societal income, so that societal income may be written as

$$I = I_0 - C(A), \tag{16.15}$$

where I_0 is (maximum) societal income in the absence of any conservation effort. Equation (16.15) can be rearranged to

$$I_0 = I + C(A) \equiv F(I, A). \tag{16.16}$$

Function $F(I,A)$ represents a curve that contains all the combinations of income I and biodiversity level A that obey the production constraint $F(I,A) = I_0$. It can be interpreted as a PPF. For the example of Eq. (16.13) and an exemplary value of $I_0 = 10$, Eq. (16.16) reads

$$I + 0.05A + 0.1A^2 = 10 \tag{16.16'}$$

and is shown in Fig. 16.3 (dashed line; note the similarity with the cost function (short-dashed line) in Fig. 9.1a when mirrored on the horizontal I-axis).

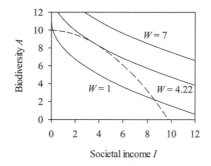

Figure 16.3 Production possibility function $F(I,A) = I_0$ after Eq. (16.16') (dashed line), and indifference function $G(I,A) = W + I_0$ after Eq. (16.18') (solid lines) for various levels of W.

With Eq. (16.15), social welfare (Eq. 9.1) can be written as

$$W \equiv V(A) - C(A) = V(A) + I - I_0 \qquad (16.17)$$

which can be rearranged to

$$W + I_0 = V(A) + I \equiv G(I, A). \qquad (16.18)$$

Function $G(I,A)$ contains all the combinations of income I and biodiversity A that generate the same welfare W. It can be interpreted as an IC like the ones in Fig. 16.2. For the example of Eq. (16.14) with $I_0 = 10$, Eq. (16.18) reads

$$I + 3A^{0.5} = W + 10 \qquad (16.18')$$

and is shown for various values of W in Fig. 16.3 (solid lines; note the similarity with the benefit function [solid line] in Fig. 9.1a when mirrored on the horizontal I-axis). Equations (16.15)–(16.18) as well as Figs 9.1 and 16.3 demonstrate the equivalence of the definition of the decision problem via cost and benefit functions on the one hand and via PPF and IC on the other.

Now turn to the socially efficient level of biodiversity A. Given PPF and IC, the task is to identify the combination of I and A that maximises W of Eq. (16.18) under the constraint of Eq. (16.16). In Section 16.4 it was shown that the optimal combination of I and A is the point on the PPF that leads to the IC associated with the highest possible utility. Graphically it is the point where IC and PPF are in contact ($I \approx 3.67$ with resulting $W \approx 4.22$ in Fig. 16.3) and tangent to each other, so that they have identical slopes.

Formally, the slope of the PPF is the derivative of A with respect to I under the constraint that $F(I,A)$ is constant. Analogously, the slope of the

IC is the derivative of A with respect to I under the constraint that $G(I,A)$ is constant. Identity of the two slopes means that

$$\left.\frac{\partial I}{\partial A}\right|_{F-const.} = \left.\frac{\partial I}{\partial A}\right|_{G=const.} . \tag{16.19}$$

Constancy of F and G implies that their total differentials, dF and dG, are zero:

$$dF = \frac{\partial F}{\partial I} dI + \frac{\partial F}{\partial A} dA = 0$$
$$dG = \frac{\partial G}{\partial I} dI + \frac{\partial G}{\partial A} dA = 0 \tag{16.20}$$

which with Eqs (16.16) and (16.18) becomes

$$dI + \frac{\partial C}{\partial A} dA = 0$$
$$dI + \frac{\partial V}{\partial A} dA = 0 . \tag{16.21}$$

With Eq. (16.19) the two equations of Eq. (16.21) are equivalent to

$$\frac{\partial V}{\partial A} = \frac{\partial C}{\partial A}, \tag{16.22}$$

which proves that the solution of the decision problem through PPF and IC is identical to the solution obtained by maximising the difference between benefit and cost.

Equation (16.22) can also be obtained formally and directly from Eqs (16.16) and (16.18). As noted, the task is to maximise W of Eq. (16.18) under the constraint of Eq. (16.16). The optimal biodiversity level can be determined by Lagrange optimisation with the Lagrangian

$$L = W - \lambda (F(I, A) - I_0) \tag{16.23}$$

and the derived equations (using Eqs (16.16) and (16.18))

$$0 = \frac{\partial L}{\partial A} = \frac{\partial W}{\partial A} - \lambda \frac{\partial F}{\partial A} = \frac{\partial V}{\partial A} - \lambda \frac{\partial C}{\partial A}$$
$$0 = \frac{\partial L}{\partial I} = \frac{\partial W}{\partial I} - \lambda \frac{\partial F}{\partial I} = 1 - \lambda \tag{16.24}$$
$$0 = \frac{\partial L}{\partial \lambda} = F(I, A) - I_0$$

The second equation on Eq. (16.24) implies $\lambda = 1$. Inserting this into the first equation delivers $\partial V / \partial A = \partial C / \partial A$. The third equation is the production constraint that had been derived from Eq. (16.15) and simply states that each euro spent for biodiversity conservation reduces the monetary societal income by one euro.

To summarise, the cost function that gives the cost of biodiversity conservation is equivalent to the PPF, and the benefit function that measures the benefits drawn from biodiversity is equivalent to the IC. The efficient level of biodiversity is found by maximising the difference between benefit and cost, which is the point at which the slopes of the benefit and cost functions are identical, while within the framework of PPF and IC, the efficient level of biodiversity is found by equating the slopes of PPF and IC. These two approaches are mathematically identical.

This equivalence has some consequences for the discussion of which of the two valuation methods, contingent valuation and choice experiment, rooted in neoclassical economics and multi-criteria analysis, more related to ecological economics, is preferable. In Section 16.4 I argued that multi-criteria analysis basically consists of the definition of a feasible set of alternatives, constrained by a PPF, which is evaluated along a set of criteria. The decision-maker(s)' preferences are measured by the IC that contains all combinations of criteria levels that are equally preferable. Thus from a formal, mathematical point of view, the two approaches of economic valuation are equivalent. This is not really surprising, since both approaches essentially require an aggregation of criteria to some joint measure. In contingent valuation and choice experiments this joint measure is money, while in multi-criteria analysis it is a score, value or whatever term is used. The differences between the approaches lie more in the terminology, in the way that trade-offs between the different criteria are dealt with, in how weights and preferences are elicited from the decision-maker(s) and in the methods used for the aggregation of the criteria.

The choice of the adequate approach, however, is not arbitrary but depends, among others, on the structure and the context of the valuation problem. For instance, multi-criteria analysis often involves a large number of criteria. A possible weighting procedure here is to ask the decision-makers to distribute one hundred small items, such as white beans, among all criteria, so that the number of items allocated to a criterion measures the relative importance of that criterion (e.g. Proctor and Drechsler 2006). Such a valuation task lies within the cognitive abilities of the decision-makers even if the number of criteria is large. In the more complex task of ranking bundles of criteria performances (attribute levels) within a choice experiment, in contrast, the number of criteria should not exceed ten, because a higher number would exceed the cognitive abilities of the respondents (DeShazo and Fermo 2002; Mangham et al. 2009). On the other hand, multi-criteria analysis usually involves a rather small group of decision-makers that should represent society (e.g. Proctor and Drechsler 2006). An advantage of such a setting is that the decision-makers can discuss and deliberate to reach a good decision. However, if the preferences of all or at least a very large number of citizens should be considered explicitly, an online choice experiment as in Drechsler et al. (2017) is probably more appropriate. To generalise, like mathematical models, all valuation approaches have their pros and cons and in each particular situation the most suitable method should be chosen with care.

Next to the choice of the adequate valuation method, the most critical part of a valuation study is the framing of the valuation problem. A sensible outcome can be expected from a valuation study only if the participants fully understand the context and the criteria of the valuation problem, the impacts of the alternatives on the criteria, the trade-offs between the criteria and the aggregation method that eventually ranks the alternatives. Economic valuation is therefore an art in and of itself and the topic of a vast number of journal articles and textbooks. The considerations in the present chapter were able to address only a few general mathematical aspects.

16.7 Natural Capital, the Economics of Biodiversity and Conclusions

Much of what has been said in the previous sections can also be considered within the concept of natural capital, which was mentioned in Section 14.1 as one of the most important types of capital (or even the most important type if its level is critically low) that constitute human survival and well-being.

Capital accounting is a major tool for the assessment of the state and development of an economy. Traditionally it has considered only the goods and services produced in an economy (the gross domestic product, GDP), but in recent decades so-called satellite accounts have been established in various countries to measure impacts on the environment (Edwards-Jones et al. 2000, ch. 11.4). A further development towards a more holistic picture of the state of the world's environmental-economic system is the index of sustainable economic welfare (ISEW) created by Daly and Cobb (1989). This index contains and aggregates various economic, social and environmental indicators, including natural resource availability, damage to the environment and others.

An attempt to measure the value of the world's natural capital in monetary units has been carried out by Costanza et al. (1997). Their study is based on the concept of ecosystem services. Ecosystem services comprise the flow of all goods and services nature provides to us humans, including so-called provisioning services such as food, fibre and fuel, regulating services such as climate and water regulation, supporting services such as nutrient cycling and soil formation, and cultural services such as spiritual, aesthetic and educational values (Millennium Ecosystem Assessment 2005). Costanza et al. (1997) compiled previous studies on a large number of ecosystem services and extrapolated them to a global scale to obtain an average value of about $33 \cdot 10^{12}$ US\$ per year.

The results of the Costanza study can be challenged for a number of conceptual and practical weaknesses, many of which were already acknowledged by the authors themselves in their paper, and the authors emphasise that their estimate represents only a rough and minimum bound on the true value of the world's ecosystem services (which actually is infinite, since many of the services are absolutely scarce (cf. Section 14.3; Baumgärtner et al. 2006), so humankind cannot survive without them).

Despite its weaknesses, the Costanza study initiated intensive debates about ecosystem services, their state and the change of their state, which eventually led to the Millennium Ecosystem Assessment MEA (Millennium Ecosystem Assessment 2005). While the MEA took mainly a natural science perspective by measuring the state of the world's environment on physical scales, a further study, TEEB ('The Economics of the Environment and Biodiversity'), tried to broaden the MEA's perspective to include social sciences, in particular economics (TEEB 2010a), as well. The TEEB study may have been motivated by the famous Stern Review (Stern 2007) on the economics of climate change and had the following main objectives (TEEB 2010a; Sukhdev et al. 2014): to

• make nature's value visible,
• contribute to a better accounting of the economy's impacts on ecosystems,
• demonstrate the value provided by protected areas,

- address the role of economic instruments for the conservation of ecosystem services,
- point to the relationship between the conservation of ecosystem services and poverty reduction, and
- altogether mainstream the economics of ecosystems, that is, help incorporate the conservation of biodiversity and ecosystem services into political and economic decision-making.

Conclusions drawn from the TEEB study include, among others, that it is important to base analyses on thorough measurement and monitoring of the ecosystem services along physical dimensions, consider the trade-offs between the conservation of different ecosystem services and the trade-offs between those ecosystem services and other socioeconomic goals, analyse these trade-offs in a spatially explicit and dynamic manner, address in particular the trade-offs between present and future values, account for risk and uncertainties, and involve the stakeholders who rely on and affect the ecosystem services (TEEB 2010a,b).

The TEEB study, as well as other economic valuation studies, has been confronted with criticism from various sides, much of which, however, may be explained by misconceptions about what economic valuation is. Sukhdev et al. (2014) try to make clear that valuation (i) is a 'human institution' that takes into account different worldviews and perceptions, rather than being an econometric exercise, (ii) is not about pricing nature or even commodifying or privatising it and (iii) is not an end in itself but a means for better communicating the importance of nature. The authors argue that without economic valuation, political and economic decision-makers tend to assign nature a price of zero and ignore it altogether.

Some people hesitate to make choices at all if these require making compromises on one objective to improve another. This is especially true if the choice has a moral dimension such as the survival of a species, or the survival and well-being of humans. However, such trade-offs are ubiquitous in almost any practical decision under constraints such as limited financial budgets, and the only choice one has is to either address these trade-offs explicitly through some form of valuation or do it ad hoc, which is unlikely to lead to a transparent and reproducible decision (Costanza et al. 1997).

To demonstrate that many decisions are choices over values, consider Figs 16.1 and 16.2a where different levels of some ecological and some economic output can be achieved through different management alternatives. As argued, all the points below the PPF represent inefficient alternatives, so that each of the two outputs can be increased without reducing the other.

Consequently, only the PPF is worth considering. In the example of Fig. 16.2a, an alternative on the PPF that leads to a rather large economic output of $x \approx 1.25$ and a medium ecological output of $y \approx 1.11$ is rational and consistent only if the economic output is given a higher (factor of 2) weight than the ecological output. Alternatively, an option that leads to a high ecological output is rational and consistent only if the ecological output is given a higher weight than the economic output. Of course, one could also choose such an alternative under a high preference for economic output. Consider, for example, the alternative that leads to a rather low economic output of $x = 0.58$ and a rather high ecological output of $y = 1.84$ and is located on the PPF. Assuming the utility function in Fig. 16.2 that assigns economic output twice the weight of ecological output, the utility level associated with $(x, y) = (0.58, 1.84)$ is about 3.0 (the upper dashed line). This is far below the utility level of 3.61 that could be achieved for that utility function under the production constraint of Eq. (16.11) – which is obtained, as shown, by the optimal choice of $x^* \approx 1.25$ and $y^* \approx 1.1$. Therefore the choice of $(x, y) = (0.58, 1.84)$ with high ecological and low economic output is not consistent with the utility function in Fig. 16.2a that gives economic output a higher weight than ecological output.

To conclude this chapter, the integration of ecology and economics has two dimensions: the supply side, which is represented by a supply or cost function or a production possibility frontier and measures what can be achieved at best under the given constraints; and the demand side, which is represented by a demand function or indifference curve and represents the values of the decision maker(s). TEEB and other valuation studies mainly focus on the demand side, while the present book mainly focuses on the supply side. Joining both sides allows us to identify efficient, that is, welfare-maximising, policy instruments and management alternatives. If knowledge about the demand side is absent and only the supply side is considered, one can still identify cost- or budget-effective policies and management alternatives that maximise an ecological output for given economic costs or budgets.

For further reading, related textbooks include Voinov (2008) (with a focus on the modelling of complex environmental systems) and Clark (2010), as well as Nunes et al. (2003), Ninan (2010) and Managi (2012) (the latter three focusing on the valuation of biodiversity and ecosystems services). Recent research articles related to Chapter 16 include

Armsworth et al. (2012), Hanley et al. (2012), Uehara (2013), Doyen et al. (2013), Akter et al. (2014), Cordier et al. (2014), Schulze et al. (2016), Turner et al. (2016), Uehara et al. (2016), Conrad and Lopes (2017), Grashof-Bokdam et al. (2017), Armsworth (2018), Brock and Xepapadeas (2018), Dislich et al. (2018), Koh et al. (2018) and Scheiter et al. (2019).

17 · *Examples of Ecological-Economic Modelling*

17.1 Superposing Ecological and Economic Solutions and the Added Value of Integration

In Section 15.1 two cases were described in which ecological-economic modelling is useful. The first is where good integrated management solutions cannot be obtained by analysing two separate disciplinary models and joining the disciplinary results, and the second is when feedback loops govern the joint ecological-economic dynamics. The present chapter presents two recent examples in more detail that demonstrate these two types of conservation problems.

To start with the first type of conservation problems, the first example joins insights from Chapters 8 and 13 and asks how the agglomeration bonus in a conservation offset scheme affects the regional coexistence of two competing species. Surun and Drechsler (2018) developed an ecological-economic model to analyse this question. In their model the agglomeration bonus is implemented as in Eq. (13.1) of Section 13.2, which states that the conservation of a land parcel produces some base level of credits of magnitude 1, plus a bonus wq that depends on the proportion q of conserved land parcels in the (eight-cell) Moore neighbourhood, multiplied by some weight w.

To initiate trade in the credits market, each landowner is obliged by the conservation agency to generate a certain amount of credits. The maximum amount that can ever be produced for given weight w is $1 + w$, which is obtained if the focal grid cell is conserved and all land parcels in the Moore neighbourhood are conserved as well ($q = 1$). Each landowner is obliged to generate a fraction $\lambda \in [0,1]$ of that maximum amount, where a value close to zero represents a case of little conservation obligation and $\lambda = 1$ means that all land parcels must be conserved. By varying the values of the policy parameters λ and w, the conservation agency can tune the landscape between a high and a low proportion of conserved land parcels and between high and low spatial agglomeration.

Next to the influence of w on the level of spatial agglomeration, a second dimension of the land-use dynamics (not mentioned in Section 13.2) is the temporal development of the land-use pattern. Since in the models of Drechsler and Wätzold (2009) and Surun and Drechsler (2018) the conservation cost on each land parcel is assumed to vary randomly in time, a profit-maximising landowner will in each time step reconsider their land-use choice and may after a strong cost increase turn a currently conserved land parcel to economic use (since at high conservation cost it is cheaper to buy credits than to keep conserving) and after a strong cost decline turn a currently economically used land parcel to conservation (since at low conservation cost it is more profitable to conserve the land and sell the earned credits). As Drechsler and Wätzold (2009) found, there is a high correlation between the temporal dynamics and the spatial pattern, so that turnover between conservation and economic use is high (low) if the level of spatial agglomeration is low (high). In other words, dispersed patterns of conservation effort are associated with high levels of disturbance and agglomerated ones are rather undisturbed.

Spatial structure and disturbance were identified in Chapter 8 as important determinants for the coexistence of species in a region. Spatial structure allows an inferior but mobile competitor to escape from a superior but less mobile competitor (cf. Section 8.4): thanks to its high mobility it will always find some land parcels unoccupied by the superior competitor.

A similar effect is generated if the habitat patches are subject to disturbance so that now and then their suitability for the species is reduced or even set to zero and after that it recovers at a certain rate (Section 8.2). If the disturbance rate is small the superior competitor may eventually outcompete the inferior competitor in the entire landscape, while at high disturbance rates the superior but less mobile competitor will not be able to reach recovered habitat patches in time and will go extinct, so that only the inferior but mobile competitor survives. At medium disturbance rates both species can coexist because the superior species can track the habitat dynamics and survive, but it cannot colonise enough habitat patches to completely replace the inferior competitor in the landscape (cf. Sections 8.6 and 8.7).

Superposing these insights from ecological modelling and economic modelling would suggest that medium levels of the agglomeration bonus (w) that generate intermediate levels of disturbance and spatial agglomeration maximise the coexistence probability of two competing species. This hypothesis has been tested by Surun and Drechsler (2018) by combining the previously described model of a conservation offset scheme with an ecological metapopulation model for two species (cf. Sections 6.1 and 8.4). In the ecological model, the superior competitor is assumed to locally replace the inferior competitor if it colonises a conserved land parcel occupied by the inferior competitor, while the inferior competitor can never colonise a conserved land parcel occupied by the superior competitor.

While both species are assumed to have the same dispersal range and can colonise only neighbouring land parcels, differences in species mobility are implicitly included in differing ratios of colonisation (c) and extinction rates (e): a species with a high ratio c/e is able to colonise many habitat patches during the expected life time $1/e$ of a local population, while a species with low c/e can colonise only a few.

The integrated ecological-economic model of Surun and Drechsler (2018) confirms the hypotheses that species coexistence is maximised if the inferior competitor has a higher mobility than the superior competitor and the disturbance rate (i.e. the agglomeration bonus) is at an intermediate level.

The explanation of these findings, however, is more complicated than suggested by the simple superposition of ecological and economic insights. Firstly, Surun and Drechsler (2018) found that at high disturbance rates (low w) even the inferior but mobile species has difficulty surviving in the landscape. Thus, how is coexistence possible if the inferior species can survive neither in the presence of the superior species at low disturbance rates nor alone at high disturbance rates? The reason is that at intermediate disturbance rates ($w \approx 0.3$) the conserved land parcels are partially agglomerated into a number of clusters. In the core of a cluster each conserved land parcel is completely surrounded by conserved land parcels and there is no turnover between conservation and economic use, that is, the disturbance rate is zero (Surun and Drechsler 2018, Fig. A1). Here the superior competitor fully outcompetes the inferior competitor. At the edges of each cluster, in contrast, the conserved land parcels have fewer conserved neighbours and are subject to moderate turnover. This prevents the superior competitor from invading the edges and leaves space for the inferior competitor.

Therefore, other than assumed in the ecological models of biodiversity, the two competitors do not inhabit the entire landscape with the inferior competitor permanently escaping from the superior competitor, but the landscape is partitioned into a subregion that is suitable for the superior competitor, and a region that is unsuitable for the superior competitor and can be inhabited by the inferior competitor.

17.2 Feedback Loops in Output-Based Payments

The second type of conservation problem that requires ecological-economic modelling is when there is a feedback loop between the ecological and the economic subsystems. Next to the examples given in Sections 15.1 and 16.3, a feedback loop can arise when a payment scheme rewards conservation outcomes rather than conservation efforts. These schemes are called output- or results-based payments, as compared with the input- or action-based payments that reward conservation efforts.

Drechsler (2017a) considered the conservation of a metapopulation in a grid-based model landscape in which each grid cell represents a land parcel that can be conserved or used for agriculture. Two payment schemes were compared with regard to their cost- and budget-effectiveness: an input-based payment where landowners receive a payment if they conserve their land parcel, and an output-based payment that rewards the presence of the target species on the land parcel.

In the input-based payment the land-use is simple and static: land parcels i that generate agricultural profits z_i below the payment p are conserved, and land parcels with higher agricultural profits are used for agriculture (cf. Section 9.3). In the model, profits z_i are constant in time, uncorrelated among land parcels and drawn from a uniform distribution with bounds $1 - \sigma$ and $1 + \sigma$. Each conserved land parcel represents a habitat patch for a local population of the metapopulation.

The output-based payment rewards the presence of a local population of the target species through a payment p. It generates more complicated land-use dynamics because there is a feedback in that the land use affects the dynamics of the metapopulation while the metapopulation dynamics affect the land-use decision; the latter is because whether conservation of a land parcel is profitable or not depends on the presence of the species.

In their decision on whether or not to conserve the land parcel, the landowner has to weigh the profit of agriculture, z_i, against the expected profit from conservation (dependent on whether the species will be present on the land parcel and the payment p). If the species happens to be present, the profit from conservation is given by the payment p, while in the absence of the species it is zero.

Altogether, the profits for the four possible states of land parcel i (defined by whether the land parcel is used for conservation or agriculture and whether it is occupied by the species or empty) are

$$\begin{aligned} \pi_i(\text{cons, occ}) &= p \\ \pi_i(\text{cons, empty}) &= 0 \\ \pi_i(\text{agr, occ}) &= p + z_i \\ \pi_i(\text{agr, empty}) &= z_i \end{aligned} \tag{17.1}$$

Denoting J_i as the probability of the species being present on the land parcel, the expected profits of conservation and agriculture are, respectively:

$$\begin{aligned} \mathrm{E}\pi_i(\text{cons}) &= J_i\pi_i(\text{cons, occ}) + (1 - J_i)\pi_i(\text{cons, empty}) = J_i p \\ \mathrm{E}\pi_i(\text{agr}) &= J_i\pi_i(\text{agr, occ}) + (1 - J_i)\pi_i(\text{agr, empty}) = J_i(p + z_i) + (1 - J_i)z_i, \end{aligned} \tag{17.2}$$

where the first term in each equation represents the case of the land parcel being occupied by a local population and the second term represents the case of the land parcel being empty (note the slight difference in the notation to eqs (1) and (2) in Drechsler (2017a), which reflects a difference in the benchmark against which profit is measured).

A risk-neutral landowner will conserve their land parcel if the expected profit $\mathrm{E}\pi$ from conservation exceeds that from agriculture and use the land parcel for agriculture otherwise. As described in Section 12.2, people are usually risk-averse, and risk aversion can be considered by mapping profit z_i into a risk-utility $u(z_i)$ (cf. Fig. 12.3). Thus Eq. (17.2) is replaced by

$$\begin{aligned} U_i(\text{cons}) &= J_i u(\pi_i(\text{cons, occ})) + (1 - J_i)u(\pi_i(\text{cons, empty})) = J_i u(p) \\ U_i(\text{agr}) &= J_i u(\pi_i(\text{agr, occ})) + (1 - J_i)u(\pi_i(\text{agr, empty})) = J_i u(p + z_i) \\ &\quad + (1 - J_i)u(z_i) \end{aligned} \tag{17.3}$$

with

$$u(x) = \frac{x^{1-\rho} - 1}{1 - \rho} \tag{17.4}$$

(cf. Eq. 12.11), and the landowner will conserve their land parcel if the expected risk-utility $U_i(\text{cons})$ from conservation exceeds that from agriculture, $U_i(\text{agr})$, and use the land parcel for agriculture otherwise.

This concludes the outline of the economic subsystem. For the ecological subsystem, Drechsler (2017a) assumed that local populations on conserved land parcels go extinct at some rate (probability per time step) e, while an empty land parcel is colonised at some rate C (cf. Section 6.1). This rate depends on the number of occupied land parcels, n, in the neighbourhood around the focal land parcel. As possible neighbourhoods,

Drechsler (2017a) considered either the eight adjacent land parcels (Moore neighbourhood) ($D = 1$), these land parcels plus their 16 adjacent neighbours ($D = 2$) or these 24 land parcels plus their 24 adjacent neighbours ($D = 3$).

If n land parcels in the neighbourhood are occupied, the probability of the focal land parcel *not* becoming colonised equals $(1 - c)^n$, where c is the probability that a local population colonises a neighbouring land parcel. The probability (per time step) of the focal land parcel *becoming* colonised then is

$$C_i = 1 - (1 - c)^n. \tag{17.5}$$

Together with the local extinction rate e, this equation defines the metapopulation dynamics on the conserved land parcels. During these dynamics the probability of a particular conserved land parcel being occupied by a local population can be shown to be

$$J_i = \frac{C_i}{C_i + e} \tag{17.6}$$

(e.g. Hanski 1994), which, via Eq. (17.3), provides the feedback from the ecological to the economic subsystem. While Drechsler (2017a) allowed for spatial spillovers such that even agricultural land parcels could be colonised and occupied with some small probability, in the following analyses this possibility is, for simplicity, not considered, so agricultural land parcels are always empty. In other words, the state (agr,occ) considered in Eqs (17.1)–(17.3) can never occur.

As boundary conditions of the simulation, Drechsler (2017a) assumed that a proportion of 5 per cent of all land parcels is always reserved and occupied by a local population (this was because if the species was absent from the entire landscape no landowner would ever expect it to become present on their land parcel and invest in conservation). All other land parcels were used for agriculture initially and in all later time steps could be conserved or used for agriculture. Each simulation step consisted of an update of the land use according to Eq. (17.3). Based on the new land-use pattern, the extinction of local populations and the colonisation of empty conserved land parcels were simulated as described previously. This generated new occupancy estimates J_i for the land parcels, which were the basis for the land-use change in the following simulation time step.

In the model analysis, Drechsler (2017a) focused on the comparison of this output-based payment with the input-based payment described previously and found that except in cases where there are high levels of risk aversion in the landowners, the output-based payment is both more cost- and budget-effective than the input-based payment. Section 17.3 will analyse the feedback loop generated by the output-based payment in more detail, while Section 17.4 will take a closer look at the associated stochasticity in the ecological-economic system and scrutinise Drechsler's (2017a) conclusion that the output-based payment is more cost-effective than the input-based payment.

17.3 Analysis of the Feedback Loop in the Output–Based Payment

As a first result of the analysis, Fig. 17.1a shows, for a particular parameterisation of the previous model of the output-based payment, the dynamics of the proportions of conserved and occupied land parcels, N_c and N_o, respectively. Although both trajectories contain considerable stochasticity, a few patterns are evident: the proportion N_c generally exceeds that of N_o, and both quantities appear to be positively correlated. These two observations are confirmed in the phase diagram of Fig. 17.1b. The correlation is not perfect but the trajectory in Fig. 17.1b is quite confined along a line with a slope of about 0.4, so generally $N_o \approx 0.4\, N_c$ and large deviations with large N_c and small N_o or vice versa are rare.

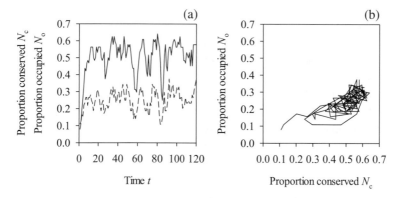

Figure 17.1 Proportions of conserved and occupied land parcels, N_c (solid line) and N_o, (dashed line) over time t (panel a), and N_o plotted versus N_c for each point in time, $t = 0, \ldots, 120$ (panel b). The trajectories represent a typical sample from a simulation run. The simulation uses the baseline parameter values of Drechsler (2017a): $\sigma = 0.5$, $\rho = 0$, $e = 0.5$, $c = 0.15$, $D = 2$ and a payment of $p = 2$.

Next to these global patterns, the question arises how the two quantities N_c and N_o interact with each other. An effective approach for this type of analysis is to fit the simulated dynamics by a Markov process (cf. Section 5.4). For this I sort all possible combinations of N_c and N_o into 20 by 20 states S_{ij}, where i and j range from 0.05 to 1.0 in steps of 0.05 and S_{ij} contains all combinations of N_c and N_o where N_c lies in the interval $(i-0.05, i]$ and N_o lies in the interval $(j-0.05, j]$. For instance, if the proportion of conserved land parcels N_c lies between 0.35 and 0.4 and the proportion of occupied land parcels N_o lies between 0.15 and 0.2, the system is in state $S_{0.4,0.2}$.

Simulating the dynamics of N_c and N_o with the ecological-economic model outlined in Section 17.2, one can record two quantities: the number of time steps at which the system resides in a particular state S_{ij}, and the frequencies of transitions between some state S_{ij} and some other state $S_{i'j'}$ $(i,j,i',j' \in \{0.05, 0.1, \ldots, 1.0\})$. To obtain significant statistics, the ecological-economic model is simulated 10,000 times over 10,000 time steps. Dividing by the number of time steps and replicates (10^8) delivers the relative frequencies F_{ij} of the states S_{ij}, and the (relative) transition frequencies $f_{iji'j'}$ between states S_{ij} and $S_{i'j'}$. The results are shown in Figs 17.2 and 17.3.

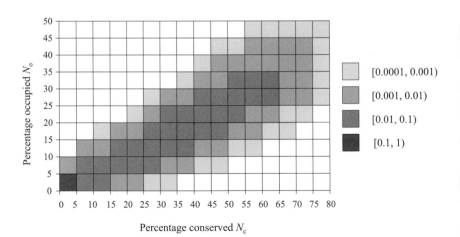

Figure 17.2 Relative frequencies of the system states. The horizontal and vertical axes measure the percentages of conserved occupied land parcels respectively, so each square in the diagram represents one out of the 20 by 20 possible states S_{ij} defined in the text. The grey scale represents the relative frequencies F_{ij} of the states S_{ij}. Model parameter values are as in Fig. 17.1.

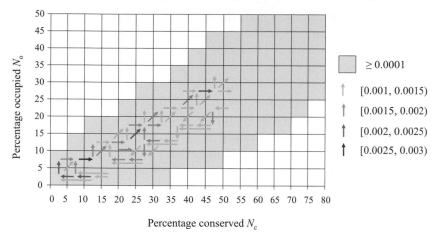

Figure 17.3 Relative frequencies of the transitions. The horizontal and vertical axes measure the percentages of conserved and occupied land parcels, respectively, so each square in the diagram represents one out of the 20 by 20 possible states S_{ij} defined in the text. The light grey colour marks states with relative frequencies $F_{ij} \geq 0.0001$. The arrows represent the transitions between the states, with their grey scale representing the relative frequencies $f_{iji'j'}$ of these transitions. Rare transitions with $f_{iji'j'} < 0.001$ are not shown for clarity. Model parameter values are as in Fig. 17.1.

In Fig. 17.2 one can see that the system is most often found in states with rather low N_c and N_o, so that combinations of $N_c \leq 0.05$ and $N_o \leq 0.05$ occur with the highest relative frequency while large $N_c > 0.8$ and $N_o > 0.5$ occur only with relative frequencies below 0.0001. For given N_c, occupancy values N_o are most likely in the range around $0.4N_c$, confirming the finding in Fig. 17.1.

Transitions with significant frequencies occur – as expected – mainly between the most frequent states (Fig. 17.3). A look at the arrows which represent the directions of the transitions reveals that for states with higher levels of occupancy N_o (larger than about $0.4N_c$) the arrows point from left to right, indicating transitions towards states with larger levels of conservation (N_c), while the opposite is observed for states with lower occupancy. Regarding the vertical arrows, some of them point up and some point down, indicating that transitions can occur towards both larger and lower occupancy.

The explanation of these observations is as follows. At high occupancy levels N_o the probability of a conserved land parcel becoming (or staying) occupied is rather large, so conservation is very likely to generate the reward $\pi(\text{cons,occ}) = p$ (cf. Eq. 17.1) which, for sufficiently high payment p, exceeds the agricultural profit z. As a consequence, landowners tend to turn agricultural land parcels into conserved ones. In contrast, at low occupancy levels a conserved land parcel is likely to stay or become empty, implying a payment of zero, which is smaller than the agricultural profit z. Consequently, landowners tend to turn conserved land parcels into agricultural ones. Altogether, for large levels of occupancy the level of conservation tends to increase while for small levels of occupancy it tends to decrease.

The impact of the level of conservation, N_c, on the transitions between different levels of occupancy, N_o, is ambiguous due to the stochasticity in the metapopulation dynamics. Although higher levels of conservation generally allow for higher levels of occupancy (cf. Fig. 17.2), even at high levels of conservation occupancy may decline to small levels. Such a transition happens when, by chance, in one or a few consecutive times steps substantially more local populations go extinct than empty land parcels are colonised.

As previously explained, such a drop in the level of occupancy is responded to with a reduction in the level of conservation – which may further reduce occupancy, and generally drives the system towards small conservation and occupancy levels, that is, close to the origin of Fig. 17.3. However, the system does not always reach such small levels of conservation and occupancy. Instead, before the conservation level has fully declined – for example at a moderate conservation level of $N_c \approx$ 0.25 – it may happen that by chance substantially more empty land parcels are colonised than local populations go extinct, so that the occupancy level increases (indicated by the upward-pointing arrows in Fig. 17.3). This is then responded to with an expansion of conservation.

Altogether the system dynamics are characterised by a feedback loop cycling clockwise. However, in contrast to the feedback loops generated by the Lotka–Volterra predator–prey model described in Section 3.5, the present feedback loop does not have a fixed centre, nor is its radius fixed nor does it follow a clear trend. Instead, both centre and radius vary randomly, so the system can run in large or small cycles around small or large levels of conservation and occupancy.

In fact, the dynamics appear almost as erratic as the dynamics generated by the discrete-time logistic growth model in Section 4.3 (Fig. 4.4e,f). However, these two types of dynamics should not be confused, since the erratic nature of the logistic growth model is due to deterministic chaos while that in the present ecological-economic system is due to stochasticity in the ecological dynamics. Moreover, the present dynamics also differ from those generated by the cobweb model in Section 2.4 and the predator–prey model in Section 3.5, because the present feedback loop is half deterministic (in the response of the level of conservation to changes in the occupancy level) and half stochastic (in the response of occupancy to conservation).

A remark for the sake of completeness: next to the drivers already described, the system dynamics are further governed and confined by an upper limit in the level of occupancy due to the 'regional carrying capacity', determined by the number of conserved land parcels, and a lower bound given by the 5 per cent of permanently reserved and occupied land parcels. And on the 'horizontal' dimension of Figs 17.1b–17.3, the number of conserved land parcels is confined by an upper limit determined by the level of the payment p, because a land parcel with high agricultural profit $z > p$ will stay in agriculture, regardless of the occupancy level (cf. Eqs (17.1)–(17.3)).

The previous analyses focused on the temporal dimension of the system dynamics and the interaction between the levels of conservation and occupancy. In the following, attention will be paid to the spatial dimension and how it couples with the temporal one. Next to the conservation and occupancy levels, N_c and N_o, variables of interest are now the proportions of conserved and occupied land parcels in the Moore neighbourhood around some focal land parcel. As focal land parcels, one may either consider all land parcels in the model landscape or only those that are conserved or occupied, respectively.

Figure 17.4a shows the dynamics of the average proportion of conserved neighbours N_{cc} around a focal land parcel compared with the proportion of conserved land parcels, N_c. One can see a rather strong correlation over time between the two variables, so that if the level of conservation N_c is large (small), so too will be the average proportion of conserved neighbours N_{cc}. This is observed regardless of whether N_{cc} is calculated on the basis of all land parcels ($N_{cc}^{(all)}$) or only for the conserved land parcels ($N_{cc}^{(c)}$). However, for the latter the correlation is slightly lower: an analysis based on 1,000 simulation runs over 1,000 times steps yields Pearson's correlation coefficients of 0.99 and 0.81, respectively. An analogous result is obtained for the relationship between the average number of occupied land parcels in the Moore neighbourhood (N_{oo}) and the proportion of occupied land parcels in the model landscape (N_o) (Fig. 17.4b).

A second observation is that the spatial variation in $N_{cc}^{(all)}$ and $N_{cc}^{(c)}$ (standard deviations of 0.21 and 0.24, respectively) is very similar to that

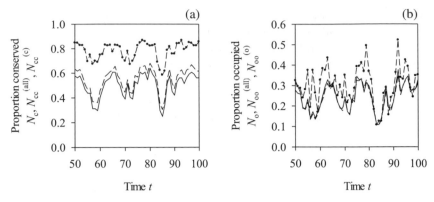

Figure 17.4 Sample trajectories of the system dynamics (subset of the time slice considered in Fig. 17.1). Panel a: proportion of conserved land parcels (N_c: solid line), average proportion of conserved land parcels in the Moore neighbourhood, considering all land parcels ($N_{cc}^{(all)}$: dashed line) and considering only conserved land parcels ($N_{cc}^{(c)}$: dashed line with circles), as functions of time t. Panel b: proportion of occupied land parcels (N_o: solid line), average proportion of occupied land parcels in the Moore neighbourhood, considering all land parcels ($N_{oo}^{(all)}$: dashed line) and considering only conserved land parcels ($N_{oo}^{(o)}$: dashed line with circles), as functions of time t. Model parameters as in Fig. 17.1.

in N_c (standard deviation of 0.21). The temporal average of $N_{cc}^{(c)}$ (0.56), however, is substantially larger than that of N_c and $N_{cc}^{(all)}$ (0.31 and 0.36, respectively). Therefore the relative variation of the conservation level around *conserved* land parcels (0.24/0.56 ≈ 0.43) is smaller than that obtained if *all* land parcels are considered (0.21/0.36 ≈ 0.58), which in turn is smaller than the relative variation in the overall level of conservation (0.21/0.31 ≈ 0.68). This implies that the relative variation in the conservation level around *conserved* land parcels is smaller than that around *empty* land parcels, indicating that the conserved land parcels form clusters in the model landscape within which the conservation level varies less strongly than outside.

This is confirmed by the above observation that $N_{cc}^{(c)}$ is less strongly correlated with N_c than $N_{cc}^{(all)}$ and allows for the conclusion that if N_c increases in the course of the model dynamics, the additional conserved land parcels generally do not appear randomly somewhere in the model landscape, but rather within gaps or at the boundary of an existing cluster of conserved land parcels. Thus, as N_c increases or declines, the clusters of conserved land parcels grow or shrink.

In contrast to the conserved land parcels, the occupied land parcels do not seem to build distinct clusters (Fig. 17.4b), which is indicated by the observation that the temporal mean of $N_{oo}^{(o)}$ only marginally exceeds the means of N_o and $N_{oo}^{(all)}$ (values of 0.21, 0.17 and 0.16, respectively). With measured standard deviations of 0.10 and 0.13, the relative variations of $N_{oo}^{(all)}$ and $N_{oo}^{(o)}$ are almost identical. An explanation for the absence of distinct clusters of the occupied land parcels is that arising clusters are likely to become disrupted soon due to the random and spatially uncorrelated (!) extinction of the local populations.

The following and final step of the analysis will address the interaction between the spatial and the temporal dimensions of the dynamics. In particular I will investigate how the number of conserved land parcels and the number of occupied land parcels in the Moore neighbourhood around some focal land parcel affect the focal land parcel's transitions between agricultural use and conservation and between the two ecological states occupied and empty. For this I define the state of a land parcel by four variables:

1. conserved ($i = 1$) or in agricultural use ($i = 0$);
2. 0–2 neighbouring land parcels conserved ($j = 1$), 3–5 neighbours conserved ($j = 2$) or 6–8 neighbours conserved ($j = 3$);
3. occupied ($k = 1$) or empty ($k = 0$); and
4. 0–2 neighbouring land parcels occupied ($l = 1$), 3–5 neighbours occupied ($l = 2$) or 6–8 neighbours occupied ($l = 3$).

Thus a land parcel altogether can assume one of $2 \times 3 \times 2 \times 3 = 36$ states S_{ijkl} (precisely, since agricultural land parcels are always empty by assumption, only 27 of these states can actually be assumed, but for simplicity and generality I also include the nine states with agricultural use and species presence, $(i,k) = (0,1)$). This implies that in total $36 \times 36 = 1,296$ transitions $S_{ijkl} \rightarrow S_{i'j'k'l'}$ from some state S_{ijkl} into some state $S_{i'j'k'l'}$ are possible. Similar to previously, I simulate the system dynamics 1,000 times with 1,000 time steps and count how often a land parcel is in a particular state S_{ijkl}, and how many transitions are of a particular type $S_{ijkl} \rightarrow S_{i'j'k'l'}$. Dividing by the total number of time steps and replicates (10^6), I obtain the relative frequencies of the states, F_{ijkl}, and the relative frequencies of the transitions, $f_{ijkli'j'k'l'}$.

I am particularly interested in how the transitions between conservation and agriculture, that is, between $i = 1$ and $i = 0$, and the transitions between occupied and empty, that is, between $k = 1$ and $k = 0$, depend on the conservation and occupancy levels in the Moore neighbourhood, that is, the levels of j and l. To start with the impact of j on i, the variable of interest is the transition rate $r_{ii'}(j)$ between states i and i' ($i, i' \in \{0,1\}$) as a function of j ($j \in \{1,2,3\}$). Since this transition may occur at any levels of k and l and may lead to any levels of j', k' and l', I take the sum over all values of these variables and calculate

$$r_{ii'}(j) = \sum_{kl j' k' l'} f_{ijkl i' j' k' l'}. \tag{17.7}$$

In an analogous manner I calculate the transition rates between conservation and agriculture as functions of the level of occupied neighbours, $r_{ii'}(l)$, the transition rates between occupied and empty as functions of the level of conserved neighbours, $r_{kk'}(j)$, and the transition rates between occupied and empty as functions of the level of occupied neighbours, $r_{kk'}(l)$. Finally I normalise the transition rates that were obtained, so that for each initial state (e.g. $i = 1$) the sum of the rate of residing in that state ($i = 1$) and the transition rate into the alternative state ($i = 0$) is one. For instance, I build

$$q_{ii'}(j) = \frac{r_{ii'}(j)}{r_{i0}(j) + r_{i1}(j)} \quad (i = 0, 1; j = 1, 2, 3) \tag{17.8}$$

which gives the proportion of transitions into states with the same land-use measure ($i' = i$) and the proportion of transitions into states with the alternative land-use measure ($i' = 1 - i$) as functions of j. The same normalisation is carried out for the other transition rates $r_{ii'}(l)$, $r_{kk'}(j)$ and $r_{kk'}(l)$.

The results are shown in Tables 17.1 and 17.2. To provide an example, the bold-faced numbers in Table 17.1 represent the rates $q_{ii'}(j = 1)$ ($i, i' \in \{0,1\}$), showing that a conserved land parcel ($i = 1$) surrounded by 0–2 conserved neighbours ($j = 1$) stays conserved ($i' = 1$) with probability 0.82 and switches to agriculture ($i' = 0$) with probability 0.18; while if the land parcel is currently in agricultural use ($i = 0$), it stays in agricultural use ($i' = 0$) with probability 0.03 and switches to conservation ($i' = 1$) with probability 0.97.

The transition rates shown in Table 17.1 allow for some clear conclusions. With increasing numbers of conserved and/or occupied neighbours (increasing j and/or l, respectively), the probability of a conserved land parcel remaining conserved ($i' = i = 1$) increases and the probability of switching to agriculture ($i' = 0$) correspondingly declines. And, with increasing numbers (j, l) of conserved and/or occupied neighbours, the probability of an agricultural land parcel remaining in agricultural use ($i' = i = 0$) declines while the probability of switching to conservation (($i' = 1$) increases (a minor, inexplicable, exception is that an increase in

Table 17.1 *Normalised transition rates* $q_{ii'}(j)$ *(column 'Conserved neighbours')* *and* $q_{ii'}(l)$ *(column 'Occupied neighbours') between Conservation (i = 1) and Agriculture (i = 0) (given by each 2×2 block) as functions of the number of conserved (j) and occupied (l) neighbours (j, l $\in$ {1,2,3}). For the bold-faced and italic numbers, see text.*

No. of neighbours		Conserved neighbours (*j*)		Occupied neighbours (*l*)	
		Conservation	Agriculture	Conservation	Agriculture
0–2	Conservation	**0.82**	**0.18**	0.88	0.12
	Agriculture	**0.03**	**0.97**	0.03	0.97
3–5	Conservation	0.92	0.08	0.98	0.02
	Agriculture	*0.07*	*0.93*	0.10	0.90
6–8	Conservation	0.95	0.05	1.00	0.00
	Agriculture	*0.06*	*0.94*	0.13	0.87

Table 17.2 *Normalised transition rates* $q_{kk'}(j)$ *(column 'Conserved neighbours') and* $q_{kk'}(l)$ *(column 'Occupied neighbours') between Occupied (k = 1) and Empty (k = 0) as functions of the number of conserved (j) and occupied (l) neighbours (j, l $\subset$ {1,2,3}). Other details as in Table 17.1.*

No. of neighbours		Conserved neighbours (*j*)		Occupied neighbours (*l*)	
		Occupied	Empty	Occupied	Empty
0–2	Occupied	0.48	0.52	0.48	0.52
	Empty	0.02	0.98	0.04	0.96
3–5	Occupied	0.48	0.52	0.49	0.51
	Empty	0.11	0.89	0.19	0.81
6–8	Occupied	0.49	0.51	0.50	0.50
	Empty	0.19	0.81	0.27	0.73

the number of conserved neighbours from 3–5 to 6–8 slightly increases the probability of an agricultural land parcel remaining in agriculture and slightly decreases its probability of becoming conserved: see the italic numbers in Table 17.1).

In an analogous manner one can conclude from Table 17.2 that with increasing numbers (*j*, *l*) of conserved and/or occupied neighbours the probability of an occupied land parcel remaining occupied (*k'* = *k* = 1) (weakly) increases and the probability of it becoming empty (*k'* = 0) correspondingly (weakly) declines; while the probability of an empty land parcel remaining empty ($k' = k = 0$) declines and the probability of it becoming occupied ($k' = 0$) increases.

In short, a large number of conserved and/or occupied land parcels in the neighbourhood increases the probability of conservation and species presence on the focal land parcel. However, next to these similarities between Tables 17.1 and 17.2, there is also an important difference. According to Table 17.1, the probability of a land parcel remaining in its current land-use type (conserved or in agricultural use, respectively) is always substantially larger than the probability of it switching to the other land-use type. In contrast, while an empty land parcel also tends to keep its state and remain empty, an occupied land parcel always has about a 50 per cent probability of becoming empty. This again reflects the stochastic nature of the metapopulation dynamics, such that local populations go extinct (at rate $e = 0.5$ in the present simulation) regardless of the processes in their neighbourhood.

All of these results can now be joined to the following description of the spatio-temporal dynamics. Starting with the initial condition of few (5 per cent) reserved and occupied land parcels, the adjacent agricultural land parcels have an increased probability of becoming conserved and occupied, so small clusters of conserved and partly occupied land parcels emerge. These continue to expand and the system moves towards higher levels of conservation and occupancy. Despite this increase, the occupied land parcels are still subject to a risk of becoming empty. If by chance several of these transitions occur in a row, the occupancy level declines, reducing the landowners' likelihoods of their land parcel being occupied, and the landowners respond by switching from conservation to agriculture. This leads to a contraction of the conserved clusters, accompanied by a further reduction of the occupancy level. The system moves back towards smaller levels of conservation and occupancy – until by chance some land parcels adjacent to (or in gaps within) a cluster of conserved and occupied land parcels become conserved and occupied, and the cycle starts anew.

17.4 The Stochasticity in the Ecological–Economic Dynamics

The stochastic cycle analysed in Section 17.3 leads to temporal fluctuations in the level of occupancy (N_o) and the total conservation cost, that is, the sum over all forgone profits z of the conserved land parcels

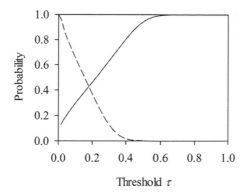

Figure 17.5 Probability of the total conservation cost lying below (solid line), and probability of the occupancy level lying above some threshold τ (dashed line) as functions of the threshold τ. Conservation cost, occupancy and threshold τ are scaled against their maximum possible values: the number of land parcels in the model landscape (note that the agricultural profits are distributed according to a uniform distribution with mean 1, so the sum of the [forgone] profits over all land parcels equals the number of land parcels).

(cf. Figs 17.1 and 17.2, considering that the total conservation cost is closely related to the number of conserved land parcels, N_c). This variation may be captured in the form of a co-viability analysis (Section 14.4) that explores the probability of the ecological-economic system staying within specified bounds. For the parameter combinations of Fig. 17.1, Fig. 17.5 shows how the probabilities of total conservation cost and occupancy level depend on the specified bounds. The probability of staying below a small cost limit is small but strongly increases with increasing cost limit, so that a cost limit of 40 per cent (measured against the maximum cost incurred if all land parcels were conserved) is met almost with certainty (Fig. 17.5, solid line). Conversely, the probability of exceeding a small occupancy target is very large, while occupancy levels above 40 per cent can never be exceeded. This type of information is helpful when assessing whether an output-based payment with the chosen level $p = 2$ is sufficiently likely to lead to an acceptable total conservation cost and an acceptable level of occupancy.

A co-viability analysis can also be used to compare the performance of input- and output-based payments. Figure 17.6 shows for both instruments the probability of the total cost staying below a specified cost limit, and the probability of the occupancy level staying above a specified occupancy target, as functions of the payment p. Not surprisingly, the probability of staying within the cost limit declines with increasing

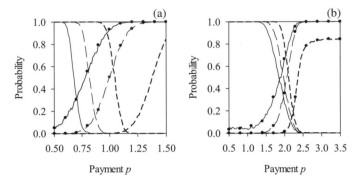

Figure 17.6 Probabilities of staying below a specified cost limit (lines without circles) and probability of staying above a specified occupancy target (lines with circles) as functions of the payment p for input-based (panel a) and output-based (panel b) payments. Cost limits: 10% (solid line), 20% (long-dashed line) and 40% (short-dashed line) relative to the maximum possible cost; occupancy targets: 10% (solid line), 20% (long-dashed line) and 40% (short-dashed line) relative to the maximum possible occupancy.

payment while the probability of exceeding the occupancy target increases.

Comparison of Figs 17.6a and b reveals that the trade-off between the probability of staying within the cost limit and the probability of exceeding the occupancy target differs between input- and output-based payments. This can be seen much more clearly in Fig. 17.7. All panels of this figure show that an increase in the probability of exceeding a specified occupancy target (henceforth termed 'ecological certainty') can be achieved only by reducing the probability of staying within the specified cost limit (henceforth termed 'economic certainty'). The marginal rate of transformation between the two quantities, represented by the negative slopes of the curves and measuring how much of the one certainty must be given up to exceed the other, is highest for low cost limits (panels a and b) and high occupancy targets (short-dashed lines). For instance, in the extreme case of the input-based payment with cost limit 0.1 (Fig. 17.7a) the short-dashed line, which represents the occupancy target of 0.4, runs entirely along the axes of the figure, meaning that for any non-zero probability of staying below the cost limit it is impossible to stay above the occupancy target.

In most cases considered in Fig. 17.7 the output-based payment (panels b,d,f) outperforms the input-based payment (panels a,c,e) in that higher ecological certainty is achieved for a given level of economic

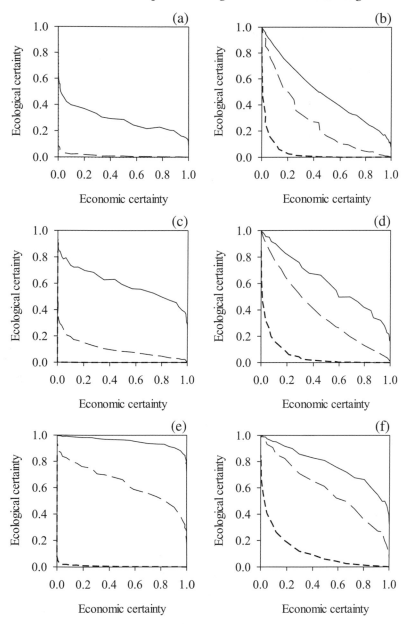

Figure 17.7 Ecological certainty (probability of occupancy level above target) as a function of economic certainty (probability of total cost below limit), obtained by varying the payment *p* from low values (implying high economic certainty and low ecological certainty) to high values (with low economic certainty and high ecological certainty). Occupancy targets: 0.1 (solid lines), 0.2 (long–dashed lines), 0.4 (short–dashed lines). Cost limits: 0.1 (panels a,b), 0.2 (panels c,d), 0.4 (panels (e,f). Input–based payment in panels a,c,e; output–based payment in panels b,d,f.

certainty, confirming Drechsler (2017a) that the output-based payment is, for the chosen model parameters, more cost-effective than the input-based payment. However, at a high cost limit (panels e,f) and low occupancy target (solid lines) the input-based payment outperforms the output-based payment.

In other words, 'weak' ecological and economic constraints can be met with higher certainty through the input-based payment, while for 'tight' constraints the output-based payment outperforms the input-based payment. This is especially true when high economic certainty (large probability of staying below the cost limit) is demanded. The reason for this is that in the input-based payment cost variation (economic uncertainty) arises only from the randomness in the agricultural profits z_i, while in the output-based payment this variation is further increased due to stochasticity in the ecological-economic dynamics.

17.5 Conclusions

The two modelling examples demonstrate the two cases mentioned in Section 15.1 in which integrated ecological-economic modelling leads to richer results than disciplinary models. The example in Section 17.1 showed that if an economic land-use model and an ecological metapopulation model are analysed systematically but separately, important conclusions may already be drawn from the verbal integration of the disciplinary model results, but some important facets such as the distribution of the two species into largely disjoint spatial clusters as a factor of regional coexistence could not be detected by the separate analyses. Even more intuitive is the conclusion that a feedback loop between an ecological and an economic subsystem, as in Sections 17.2–17.4, cannot be adequately modelled and analysed through separate disciplinary models.

Although the two examples do not cover all the features of the modelling framework by Costanza et al. (1993) described in Section 14.4, they involve already quite rich ecological-economic dynamics due to the non-linearities, stochasticity and feedback loops considered. Moreover, scale issues highlighted by Costanza et al. (1993) were shown to play a considerable role in the model of the output-based payment model where the landowners respond to the amount of land parcels occupied by the species in their neighbourhood rather than the number of occupied land parcels in the entire model landscape – which leads to the observed spatial clustering of the conserved land parcels (Section 17.3). The 'limits of predictability in modelling ecological economic systems' (Costanza et al. 1993) are obvious in the observation that the system shows feedback loops with unpredictable location and size in the N_c–N_o phase space.

Both examples, especially the second one, demonstrate that market-based instruments – although they are more a topic of neoclassical environmental economics – can lead to complex ecological-economic dynamics, calling for analysis within an ecological-economic framework. The concept of co-viability introduced by Baumgärtner and Quaas (2009) was used to derive a stochastic equivalent of the 'neoclassical' production possibility frontier (cf. Section 16.2): instead of depicting conventional trade-offs between mean values of system variables, it shows trade-offs between the probabilities of these system variables staying within specified bounds. From this experience one can argue that ecological-economic modelling not only integrates ecological and economic concepts and knowledge but can also serve as a bridge between the two subdisciplines of environmental economics and ecological economics.

The two modelling examples already present some interesting insights into the interplay of species dynamics and species interactions, as well as spatial pattern formation – be it induced by an agglomeration bonus as in Section 17.1 or emerging from the ecological-economic dynamics as in Section 17.2 – and temporal feedback loops. Extension of the models could, for instance, include more species with more types of inter-species interactions to transcend from the one- or two-species level to the community or ecosystem level, and consider a richer behaviour of the landowners by including, for example, memory, learning and inequity aversion. This is likely to generate more of the features mentioned by Costanza (1993) as elements of his 'general framework for complex ecological economic systems dynamics', such as multiple equilibria, path dependence and surprise. And finally, the conceptual models presented here might be applied and parameterised to real-world systems so they would not only serve the purpose of generating theoretical understanding but also help manage concrete systems in a sustainable manner. Altogether, a large number of concepts and approaches exist for the development and analysis of purposeful ecological-economic models. It is hoped that this book can provide some assistance and motivation for such endeavours.

18 · *Outlook*

This book tries to give an overview on the main foundations and state of the art of ecological-economic modelling. Yet the field is dynamic and ecological-economic models evolve as better hardware and software and new concepts for model development are developed and new environmental problems emerge. Five avenues of future research (there are certainly more) may be highlighted: (i) multiple interacting species, (ii) multiple ecosystem services, (iii) feedbacks into the policy level, (iv) managing dynamic complexity and (v) making models relevant for policymakers.

18.1 Multiple Interacting Species

A first challenge for future ecological-economic modelling follows from the growing body of ecological research on species interactions and (meta)communities. Although quite a few ecological-economic models (several of which are mentioned in this book) consider multiple species, the interactions between the species are usually ignored. Consequently, ecological processes on the ecosystem level cannot be addressed. In Chapter 8, however, we saw that the functioning of an ecosystem cannot be understood without taking species interactions into account. The analysis of ecological-economic models that ignore this issue therefore may lead to wrong conclusions, for example with regard to what policies are cost-effective for the conservation and sustainable management of an ecosystem.

A few general insights into the impacts of human activities and other drivers on the coexistence and abundances of the species in an ecosystem have been outlined in Chapter 8, including the intermediate disturbance hypothesis, the competition–colonisation trade-off and the role of endogenous disturbances that prevent, for instance, a savannah from drifting either into woodland or grassland. Incorporating these concepts into the development of ecological-economic models is certainly rewarding.

A few ecological-economic models exist that consider multiple interacting species. Most of them seem to consider marine systems, analysing the impact of harvesting on species abundances. Examples include Gourguet et al. (2013), Metcalfe et al. (2015), Marzloff et al. (2016) and Bertram and Quaas (2017). A notable example of a terrestrial model study is Daniels et al. (2017), who analyse the impact of natural predators on a pest species and the implied farm income in an agro-ecosystem.

All of these model studies take a regulation or planning approach in which measures are implemented through a perfectly informed fisheries manager, farm manager or similar. While the role of market-based instruments in the conservation of multiple *independent* species has been analysed in various studies (see, e.g., Polasky et al. (2008), Nelson et al. (2008), Wätzold et al. (2016) and further examples in this book), the question of how market-based conservation instruments should be designed to conserve multiple *interacting* species has not yet been addressed satisfactorily. A modest first step in this direction is the conservation offset model for two competing species presented in Section 17.1, but there are ample rewarding open research questions around this topic.

18.2 Multiple Ecosystem Services

In the present book, the conservation of biodiversity is considered a policy objective without justifying it further based on the benefits it provides to humans (cf. Millennium Ecosystem Assessment (2005) and Section 16.7). In most ecological-economic models, biodiversity (be it measured by the number of species, the survival probability of particular target species or otherwise) is an end in itself, independent of its role in providing ecosystem services such as crop pollination, water purification, carbon sequestration and so forth

A few articles consider one or more of these ecosystem services within an ecological-economic model. Two have already been mentioned: Nelson et al. (2008) consider the number of species and the amount of carbon sequestered as two, partly conflicting, ecosystem services; while in Daniels et al. (2017) three natural predators control a pest species and provide agricultural benefits. Further examples include Gutrich et al. (2016), who consider the ecosystem service of dust suppression; Cong et al. (2016), who focus on the spatial allocation of land-use measures to enhance crop pollination; and Sanga and Mungatana (2016) and Wang et al. (2017), who address surface water quality.

A major effort towards the consideration of multiple ecosystem services has been the development of the software InVEST (Integrated Valuation of Ecosystem Services and Tradeoffs), which is 'a suite of models used to map and value the goods and services from nature that sustain and fulfil human life' (naturalcapitalproject.stanford.edu/soft ware/). The ecosystem services addressed in the software include, among others, carbon sequestration, crop pollination, habitat quality, recreation and water purification. Embedding a model such as InVEST within a numerical optimisation framework allows one to derive production possibility frontiers and analyse trade-offs between the different ecosystem services (Verhagen et al. 2018).

Similar to the models for multiple interacting species, InVEST and optimisation approaches such as Verhagen et al. (2018) assume a regulator who can assign land-use measures at will to achieve desired levels of ecosystem services (under a budget constraint), but they do not address the question of how this desired land-use pattern can be implemented in a real socioeconomic environment with the imperfections discussed in Part III of this book. In particular, the role of market-based environmental instruments is grossly under-addressed in this context and a rewarding avenue for future ecological-economic research.

18.3 Feedback into the Policy Level

As demonstrated in Chapters 14–17, feedback loops between the ecological and economic subsystem are an important constituent of many ecological-economic systems and models. In the examples of Section 16.3 the feedback from the ecological to the economic subsystem arises because agents respond to changes in their environment. These agents are typically landowners, fishermen or similar. In addition to these 'lower-level' agents, the models often include an agent on a 'higher' level: the regulator or conservation agency. This agent sets policies and incentives to which the 'lower-level' agents react in the ecological-economic system. In most cases the regulator's decisions in terms of policy design are considered an exogenous factor that may be varied by the model analyst to explore alternative policies, but they are not an endogenous part of the model dynamics themselves. A feedback from the ecological subsystem and/or the landowners to the regulator is not considered.

An example of a model with two governance levels, regulator and landowners, is by Schlüter and Pahl-Wostl (2007), who consider land use in a semi-arid basin. Both the regulator, represented by a national authority, and the farmers can decide on the allocation and use of scarce water. The allocation of the water is decided on the basis of water flows observed in previous years; the farmers, in addition, also consider past farming yields. Thus, the agents on both governance levels act dynamically and endogenously within the model. However, since the land-use decisions do not influence the water flow in future years, which is the sole dynamic factor affecting the regulator's decision, there is no feedback from the farmers' governance level to the level of the national authority.

The adaptation of governance to changes in the ecological-economic system has been analysed empirically, for example by Theesfeld and McKinnon (2014). Their case of interest is a water rights system in Wyoming that was challenged by increasing calls for the adequate consideration of the rights of water birds. Through interviews, the authors reveal the attitudes of key players in the governance system towards the system's ability to cope with change and derive conclusions under which such a system can gradually adapt. While the authors' conclusions represent valuable insights into the response of a governance system to change, ecological-economic modelling has enormous potential to enhance such research due to its ability to formulate and test alternative hypotheses through simulation.

18.4 Managing Dynamic Complexity

A few cases of dynamic complexity have been considered in this book, including feedback loops (Sections 3.5, 16.3, 17.3), chaos (Section 4.3) and resilience (Sections 2.8, 14.3). A relevant feature within the resilience concept is tipping points, that is, boundaries in the system's state space which, when crossed, imply drastic changes in the structure and the functioning of the system. In Fig. 2.5 a tipping point is represented by the ridge between the two valleys: when crossed from left to right the system moves into a state with qualitatively different characteristics.

In the examples mentioned we have seen how management can affect the dynamics of a complex system. How successful a management strategy is with respect to maintaining the system in a desired state can be measured, for instance, by co-viability analysis (Sections 14.4, 17.4). In a way, such an analysis may be regarded as trial and error: a set of management strategies is proposed and the strategies are compared with each other to identify the one that maximises co-viability.

A contrasting approach that originates from optimal control theory and involves dynamic optimisation to analyse how a system should be managed optimally to maximise some long-term objective has been proposed by Fenichel and Horan (2016). While classical optimal control problems usually assume 'well-behaved' systems with rather simple dynamics, the authors investigate how an ecological-economic system with non-convexities, tipping points and the potential for complex dynamics should be managed optimally and how its path through the state space can be controlled.

Next to tipping points, a second challenge in the management of dynamic complexity is uncertainty and surprise. Global changes, such as climate and land-use changes, involve high uncertainties, and it is of great importance to develop policies and strategies to cope with these uncertain changes. A general recommendation for such policies may be to be flexible. In a changing ecological and economic environment one may ask, for example, whether a conservation agency should purchase land for conservation or keep it in private ownership and pay the landowner to carry out conservation measures (Schöttker et al. 2016). These two strategies differ in their costs and levels of flexibility, and their relative performances depend on the characteristics of the managed system.

The opposite of flexibility is lock-in and path dependence. A purchased land parcel is usually 'locked in', in that it cannot easily be sold to purchase another land parcel that has become ecologically more valuable. Path dependence is tightly related to lock-in and means that past decisions constrain today's decision space. Both issues are major challenges in times of global change. They have been addressed by a few articles such as Strange et al. (2006) and Boettiger et al. (2016), but all of them take a planning approach while the role of market-based instruments in this context has not yet been analysed.

18.5 Making Models Relevant for Policymakers

Conservation biology – and, one could add, biodiversity economics – is said to be a 'crisis discipline' that has not only an academic but also a political and a societal dimension. Although ecological-economic research offers fascinating intellectual challenges, one should always be aware that biodiversity is a sensitive and precious good that is under severe threat. Biodiversity researchers thus have a particular responsibility for their study objects, not least because these are living creatures.

This leads to the challenge of making research in the fields of bio-diversity economics and ecological-economic modelling relevant for society and policymakers. Two approaches to doing so may be high-lighted here. The first is the development of decision support software that allows decision-makers such as conservation agencies or conserva-tion non-governmental organisations to use models without having to be modellers themselves. A successful example here is the Marxan software (cf. Sections 2.7, 9.2) that has been applied to numerous reserve selection problems.

A challenge here is to develop tools that are specific enough for each case but general enough to be applicable to many similar conservation problems. Reserve selection problems are comparatively easy to standardise, possibly explaining Marxan's wide distribution. The InVEST software mentioned previously is still quite general but contains a number of complicated elements that need to be tailored to the specific problem. The software DSS-Ecopay outlined in Section 2.7 is, and needs to be, rather specific and detailed in its consideration of agricultural land-use measures, but this means that it is less easy to transfer it to other conservation problems (although a few successful examples of such a transfer of DSS-Ecopay exist).

A second, more general approach to making ecological-economic research more policy-relevant is to improve the involvement of policy-makers in the research – for instance by letting them contribute (more) to the formulation of research objectives and comment on the political feasibility and practicality of ideas developed in the research project. In its calls for research proposals, the German Federal Ministry of Education and Research, for example, explicitly requires the active involvement of stakeholders beyond a mere role as external advisors. A model for such interaction between researchers and policymakers may be the co-creation concept developed in the field of business economics that links, for example, a company with its customers to develop mutually benefi-cial outcomes. In the present case, the beneficiaries of this co-creation would be both researchers and decision-makers, as well as the society and its natural environment.

18.6 Concluding Remarks

What philosophy should guide integrated modelling efforts? Probably the most important prerequisites for successful integrated modelling are openness and communication among the involved disciplines. According to the author's own experience, communication is facilitated if the model is developed to understand or solve a concrete problem or answer a concrete question. Whether the model developed is of a general or a specific type, the focus on a concrete problem usually provides a common basis on which knowledge can be exchanged most effectively. The focus on a concrete problem further helps avoid getting trapped in ideological debates about the 'right' scientific approach. In modelling, there is no right or wrong but only useful or useless, or as Box (1979, p. 202) puts is: 'All models are wrong but some are useful.'

References

Aiello-Lammens, M.E., Akçakaya, H.R., 2016. Using global sensitivity analysis of demographic models for ecological impact assessment. *Conservation Biology* **31**, 116–25.

Akçkaya, H.R., Burgman, M.A., Ginzburg, L.R., 1999. *Applied Population Ecology: Principles and Computer Exercises*. Sinauer, 2nd ed.

Akter, S., Grafton, R.Q., Merritt, W.S., 2014. Integrated hydro-ecological and economic modelling of environmental flows: Macquarie Marshes, Australia. *Agricultural Water Management* **145**, 98–109.

Albers, H.J., Ando, A.W., Batz, M., 2008. Patterns of multi-agent land conservation: Crowding in/out, agglomeration, and policy. *Resource and Energy Economics* **30**, 492–508.

Albers, H.J., Lee, K.D., Sims, K.R.E., 2018. Economics of habitat fragmentation: A review and critique of the literature. *International Review of Environmental and Resource Economics* **11**, 97–144.

Alligood, K.T., Sauer, T.D., Yorke, J.A., 2000. *Chaos: An Introduction to Dynamical Systems*. Springer.

Anderies, J.M., 2000. On modelling human behaviour and institutions in simple ecological economic systems. *Ecological Economics* **35**, 393–412.

Ando, A.W., Camm, J., Polasky, S., Solow, A., 1998. Species distributions, land values, and efficient conservation. *Science* **279**, 2126–8.

Ando, A.W., Mallory, M.L., 2012. Optimal portfolio design to reduce climate-related conservation uncertainty in the Prairie Pothole Region. *Proceedings of the National Academy of Sciences of the USA* **109**, 6484–9.

Annachiarico, B., Di Dio, F., 2017. GHG emissions control and monetary policy. *Environmental and Resource Economics* **67**, 823–51.

Armsworth, P.R., 2018. Time discounting and the decision to protect areas that are near and threatened or remote and cheap to acquire. *Conservation Biology* **32**, 1063–73.

Armsworth, P.R., Acs, S., Dallimer, M., Gaston, K.J., Hanley, N., Wilson, P., 2012. The cost of policy simplification in conservation incentive programs. *Ecology Letters* **15**, 406–14.

Armsworth, P.R., Daily, G.C., Kareiva, P., Sanchirice, J.N., 2006. Land market feedbacks can undermine biodiversity conservation. *Proceedings of the National Academy of Sciences of the USA* **103**, 5403–8.

Asheim, G.B., 2013. A distributional argument for supply-side climate policies. *Environmental and Resource Economics* **56**, 239–54.

Atewamba, C., Nkuiya, B., 2017. Testing the assumptions and predictions of the Hotelling Model. *Environmental and Resource Economics* **66**, 169–203.

Axelrod, R., Hamilton, W.D., 1981. The evolution of cooperation. *Science* **27**, 1390–6.

Baker, C.M., Gordon, A., Bode, M., 2016. Ensemble ecosystem modeling for predicting ecosystem response to predator reintroduction. *Conservation Biology* **31**, 376–84.

Ball, I.R., Possingham, H.P., Watts, M., 2009. Marxan and relatives: Software for spatial conservation prioritisation. In Moilanen, A., Wilson, K.A., Possingham H.P. (eds.), *Spatial Conservation Prioritisation: Quantitative Methods and Computational Tools*. Oxford University Press, pp. 185–95.

Bamière, L., David, M., Vermont, B., 2013. Agri-environmental policies for biodiversity when the spatial pattern of the reserve matters. *Ecological Economics* **85**, 97–104.

Banerjee, S., Cason, T.N., de Vries, F.P., Hanley, N., 2018. Spatial coordination and joint bidding in conservation auctions. Paper presented at the BIOECON conference 2018 in Cambridge, UK. www.bioecon-network.org/pages/20th %202018/Hanley_1.pdf (last accessed 10 October 2019).

Banerjee, S., de Vries, F.P., Hanley, N., van Soest, D.P., 2014. The impact of information provision on agglomeration bonus performance: An experimental study on local networks. *American Journal of Agricultural Economics* **96**, 1009–29.

Banerjee, S., Kwasnica, A.M., Shortle, J.S., 2012. Agglomeration bonus in small and large local networks: A laboratory examination of spatial coordination. *Ecological Economics* **84**, 142–52.

Banerjee, S., Kwasnica, A.M., Shortle, J.S., 2015. Information and auction performance: A laboratory study of conservation auctions for spatially contiguous land management. *Environmental and Resource Economics* **61**, 409–31.

Barabás, G., D'Andrea, R., Stump, S.M., 2018. Chesson's coexistence theory. *Ecological Monographs* **88**, 277–303.

Barraquand, F., Louca, S., Abbott, K.C., et al., 2017. Moving forward in circles: Challenges and opportunities in modelling population cycles. *Ecology Letters* **20**, 1074–92.

Barraquand, F., Martinet, V., 2011. Biological conservation in dynamic agricultural landscapes: Effectiveness of public policies and trade-offs with agricultural production. *Ecological Economics* **70**, 910–20.

Baskett, M.L., Waples, R.S., 2012. Evaluating alternative strategies for minimizing unintended fitness consequences of cultured individuals of wild populations. *Conservation Biology* **27**, 83–94.

Bátary, P., Dicks, L.V., Kleijn, D., Sutherland, W.J., 2015. The role of agri-environment schemes and environmental management. *Conservation Biology* **29**, 1006–16.

Bauer, D.M., Wing, I.S., 2016. The macroeconomic cost of catastrophic pollinator declines. *Ecological Economics* **126**, 1–13.

Baumgärtner, S., Becker, C., Faber, M., Manstetten, R., 2006. Relative and absolute scarcity of nature: Assessing the roles of economics and ecology for biodiversity conservation. *Ecological Economics* **59**, 487–98.

Baumgärtner, S., Becker, C., Frank, K., Müller, B., Quaas, M., 2008. Relating the philosophy and practice of ecological economics: The role of concepts, models, and case studies in inter- and transdisciplinary sustainability research. *Ecological Economics* **67**, 384–93.

Baumgärtner, S., Quaas, M.F., 2009. Ecological-economic viability as a criterion of strong sustainability under uncertainty. *Ecological Economics* **68**, 2008–20.

Becher, M.A., Grimm, V., Thorbeck, P., Horn, J., Kennedy, P.J., Osborne, J.L., 2014. BEEHAVE: A systems model of honeybee colony dynamics and foraging to explore multifactorial causes of colony failure. *Journal of Applied Ecology* **51**, 470–82.

Begon, M., Townsend, C.R., Harper, J.L., 1990. *Ecology: From Individuals to Ecosystems*. Blackwell, 2nd ed.

Beissinger, S.R., McCullough, D.R. (eds.), 2002. *Population Viability Analysis*. University of Chicago Press.

Bell, A., Parkhurst, G., Droppelmann, K., Benton, T.G., 2016. Scaling up pro-environmental agricultural practice using agglomeration payments: Proof of concept from an agent-based model. *Ecological Economics* **126**, 32–41.

Ben-Haim, Y., 2001. *Information-Gap Theory: Decisions under Severe Uncertainty*. Academic Press.

Bertram, C., Quaas, M.F., 2017. Biodiversity and optimal multi-species ecosystem management. *Environmental and Resource Economics* **67**, 321–50.

Bode, M., Probert, W., Turner, W.R., Wilson, K.A., Venter, O., 2011. Conservation planning with multiple organisations and objectives. *Conservation Biology* **25**, 295–304.

Bode, M., Wilson, K.A., Brooks, T.M., et al., 2008. Cost-effective global conservation spending is robust to taxonomic group. *Proceedings of the National Academy of Sciences of the USA* **105**, 6489–501.

Boettiger, C., Bode, M., Sanchirico, J.N., et al., 2016. Optimal management of a stochastically varying population when policy adjustment is costly. *Ecological Applications* **26**, 808–17.

Bolton, G.E., Ockenfels, A., 2000. A theory of equity, reciprocity, and competition. *American Economic Review* **90**, 166–93.

Börgers, T., Krähmer, D., Strausz, R., 2015. *An Introduction to the Theory of Mechanism Design*. Oxford University Press, 1st ed.

Box, G.E.P., 1979. Robustness in scientific model building. In Launer, R.L., Wilkinson, G.N. (eds.), *Robustness in Statistics*. Academic Press, pp. 201–36.

Brans, J.P., Mareschal, B., 1990. The PROMETHEE methods for MCDM; the PROMCALC, GAIA and BANKADVISER software. In Bana e Costa, C.A. (ed.), *Readings in Multiple Criteria Decision Aid*. Springer, pp. 216–52.

Brauchli, K., Killingback, T., Doebeli, 1999. Evolution of cooperation in spatially structured populations. *Journal of Theoretical Biology* **200**, 405–17.

Brinck, K., Fischer, R., Groeneveld, J., et al., 2017. High resolution analysis of tropical forest fragmentation and its impact on the global carbon cycle. *Nature Communications* **8**, 14855.

Brock, W.A., Xepapadeas, A., 2018. Modeling coupled climate, ecosystems, and economic systems. *Handbook of Environmental Economics* **4**, 1–60.

Bull, J.W., Gordon, A., Law, E.A., Suttle, K.B., Millner-Gulland, E.J., 2014. Importance of baseline specification in evaluating conservation interventions and achieving no net loss of biodiversity. *Conservation Biology* **28**, 799–809.

Bulte, E.H., Horan, R.D., 2003. Habitat conservation, wildlife extraction and agricultural expansion. *Journal of Environmental Economics and Management* **45**, 109–27.

Burgman, M.A., Ferson, S., Akçakaya, H.R., 1993. *Risk Assessment in Conservation Biology*. Chapman & Hall.

Caswell, H., 2001. *Matrix Population Models*. Sinauer.

Chaikin, P.M., Lubensky, T.C., 1995. *Principles of Condensed Matter Physics*. Cambridge University Press.

Chavas, J.-P., Di Falco, S., 2017. Resilience, weather and dynamic adjustments in agroecosystems: The case of wheat yield in England. *Environmental and Resource Economics* **67**, 297–320.

Chen, S., Dobramysl, U., Täuber, U.C., 2018. Evolutionary dynamics and competition stabilize three-species predator–prey communities. *Ecological Complexity* **36**, 57–72.

Chesson, P., 2000. Mechanisms of maintenance of species diversity. *Annual Review of Ecology, Evolution and Systematics* **31**, 343–66.

Chobotová, V., 2013. The role of market-based instruments for biodiversity conservation in Central and Eastern Europe. *Ecological Economics* **95**, 41–50.

Choi, P.-S., Espínola-Arredondo, A., Muñoz-García, F., 2018. Conservation procurement auctions with bidirectional externalities. *Journal of Environmental Economics and Management* **92**, 559–79.

Chomitz, K.M., 2004. Transferable development rights and forest protection: An exploratory analysis. *International Regional Science Review* **27**, 348–73.

Clark, C.W., 2010. *Mathematical Bioeconomics: The Mathematics of Conservation*. Wiley.

Collard, B., Tixier, P., Carval, D., Lavigne, C., Delattre, T., 2018. Spatial organisation of habitats in agricultural plots affects per-capita predator effect on conservation biological control: An individual based modelling study. *Ecological Modelling* **388**, 124–35.

Comello, S.D., Maltais-Landry, G., Schwegler, B.R., Lepech, M.D., 2014. Firm-level ecosystem service valuation using mechanistic biogeochemical modeling and functional substitutability. *Ecological Economics* **100**, 63–73.

Common, M., Perrings, C., 1992. Towards an ecological economics of sustainability. *Ecological Economics* **6**, 7–34.

Cong, R.-G., Ekroos, J., Smith, H.G., Brady, M.V., 2016. Optimizing intermediate ecosystem services in agriculture using rules based on landscape composition and configuration indices. *Ecological Economics* **128**, 214–23.

Connell, J.H., 1978. Diversity in tropical rain forests and coral reefs. *Science* **199**, 1302–10.

Conrad, J.M., Lopes, A.A., 2017. Poaching and the dynamics of a protected species. *Resource and Energy Economics* **48**, 55–67.

Cordier, M., Pérez Agúndez, J.A., Hecq, W., Hamaide, B., 2014. A guiding framework for ecosystem services monetization in ecological-economic modeling. *Ecosystem Services* **8**, 86–96.

Costanza, R., 1989. What is ecological economics? *Ecological Economics* **1**, 1–7.

Costanza, R., d'Arge, R., de Groot, R., et al., 1997. The value of the world's ecosystem services and natural capital. *Nature* **387**, 253–60.

Costanza, R., Cumberland, J., Daly, H., Goodland, R., Noorgard, R., 1997. *An Introduction to Ecological Economics*. St. Lucie Press and ISEE.

Costanza, R., Wainger, L, Folke, C., Maler, K.-G. 1993. Modeling complex ecological economic systems: Toward an evolutionary, dynamic understanding of people and nature. *BioScience* **43**, 545–55.

Costello, C., Polasky, S., 2004. Dynamic reserve site selection. *Resource and Energy Economics* **26**, 157–74.

Coulson, T., Mace, G.M., Hudson, E., Possingham, H.P., 2001. The use and abuse of population viability analysis. *Trends in Ecology & Evolution* **16**, 219–21.

Creti, A., Kotelnikova, A., Meunier, G., Ponssard, J.-P., 2018. Defining the abatement cost in presence of learning-by-doing: Application to the fuel cell electric vehicle. *Environmental and Resource Economics* **71**, 777–800.

Cumming, G.S., 2018. A review of social dilemmas and social-ecological traps in conservation and natural resource management. *Conservation Letters* **11**, 1–15.

Dalmazzone, S., Giaccaria, S., 2014. Economic drivers of biological invasions: A worldwide, bio-geographic analysis. *Ecological Economics* **105**, 154–65.

Daly, H., Cobb, J., 1989. *For the Common Good–Redirecting the Economy towards Community, the Environment and Sustainable Development*. Green Print.

Daly, H.E., Farley, J., 2010. *Ecological Economics: Principles and Applications*. Island Press, 2nd ed.

Daniels, S., Witters, N., Beliën, T., Vrancken, K., Vangronsveld, J., Van Passel, S., 2017. Monetary valuation of natural predators for biological pest control in pear production. *Ecological Economics* **134**, 160–73.

Dantas de Paula, M., Groeneveld, J., Huth, A., 2015. Tropical forest degradation and recovery in fragmented landscapes: Simulating changes in tree community, forest hydrology and carbon balance. *Global Ecology and Conservation* **3**, 664–77.

Denman, K.L., Brasseur, G., Chidthaisong, A., et al., 2007. Couplings between changes in the climate system and biogeochemistry. In Solomon, S., Qin, D., Manning, M., et al. (eds.), *Climate Change 2007: The Physical Science Basis. Contribution of Working Group I to the Fourth Assessment Report of the Intergovernmental Panel on Climate Change*. Cambridge University Press.

Dennis, B., Assas, L., Elaydi, S., Kwessi, E., Livadiotis, G., 2016. Allee effects and resilience in stochastic populations. *Theoretical Ecology* **9**, 323–35.

Derissen, S., Quaas, M.F., 2013. Combining performance-based and action-based payments to provide environmental goods under uncertainty. *Ecological Economics* **85**, 77–84.

Derissen, S., Quaas, M.F., Baumgärtner, S., 2011. The relationship between resilience and sustainability of ecological-economic systems. *Ecological Economics* **70**, 1121–8.

DeShazo J.R, Fermo, G., 2002. Designing choice sets for stated preference methods: The effects of complexity on choice consistency. *Journal of Environmental Economics and Management* **44**, 123–43.

Deutsch, A., Dormann, S., 2005. *Cellular Automaton Modeling of Biological Pattern Formation*. Birkhäuser.

Dhanjal-Adams, K.L., Klaassen, M., Nicol, S., Possingham, H.P., Chadès, I., Fuller, R.A., 2017. Setting conservation priorities for migratory networks under uncertainty. *Conservation Biology* **31**, 646–56.

Diamantaras, D., Cardamone, E., Campbell, K., Deacle, S., Delgado, L., 2009. *A Toolbox for Economic Design: 2009 Edition*. Palgrave Macmillan.

Di Minin, E., Laitila, J., Montesino-Pouzols, F., et al., 2014. Identification of policies for a sustainable legal trade in rhinoceros horn based on population projection and socioeconomic models. *Conservation Biology* **29**, 545–55.

Dislich, C., Hettig, E., Salecker, J., Heinonen, J., Lay, J., Meyer, K.M., Wiegand, K., Tarigan, S., 2018. Land-use change in oil palm dominated tropical landscapes: An agent-based model to explore ecological and socio-economic trade-offs. *PLoS one* **13** (1), e0190506.

Dixit, A.K., Pindyck, R.S., 1994. *Investment under Uncertainty*. Princeton University Press.

Doak, D.F., Bakker, V.J., Vickers, W., 2013. Using population viability criteria to assess strategies to minimize disease threats for an endangered carnivore. *Conservation Biology* **27**, 303–14.

Dobson, A.J., 1983. *Introduction to Statistical Modelling*. Springer.

Doyen, L., Cissé, A., Gourguet, S., et al., 2013. Ecological-economic modelling for the sustainable management of biodiversity. *Computational Management Science* **10**, 353–64.

Drechsler, M., 2011. Trade-offs in the design of cost-effective habitat networks when conservation costs are variable in space and time. *Biological Conservation* **144**, 479–89.

Drechsler, M., 2017a. Performance of input- and output-based payments for the conservation of mobile species. *Ecological Economics* **134**, 49–56.

Drechsler, M., 2017b. Generating spatially optimized habitat in a trade-off between social optimality and budget efficiency. *Conservation Biology* **31**, 221–5.

Drechsler, M., 2017c. The impact of fairness on side payments and cost-effectiveness in agglomeration payments for biodiversity conservation. *Ecological Economics* **141**, 127–35.

Drechsler, M., 2018. Modelling the effectiveness and permanence of a compensation payment scheme for the conservation of a public environmental good. Presentation at the BIOECON conference 2018 in Cambridge, UK. www.bioecon-network.org/pages/20th%202018/Drechsler.pdf (last accessed 10 October 2019).

Drechsler, M., Egerer, J., Lange, M., et al., 2017. Efficient and equitable spatial allocation of renewable power plants at the country scale. *Nature Energy* **2**, 17124.

Drechsler, M., Grimm, V., Myšiak, J., Wätzold, F., 2007. Differences and similarities between ecological and economic models for biodiversity conservation. *Ecological Economics* **62**, 232–41.

Dressler, G., Groeneveld, J., Buchmann, C.M., et al., in press. Implications of behavioral change for the resilience of pastoral systems: Lessons from an agent-based model. *Ecological Complexity*.

Drechsler, M., Hartig, F., 2011. Conserving biodiversity with tradable permits under changing conservation costs and habitat restoration time lags. *Ecological Economics* **70**, 533–41.

Drechsler, M., Johst, K., Wätzold, F., Shogren, J.F., 2010. An agglomeration payment for cost-effective biodiversity conservation in spatially structured landscapes. *Resource and Energy Economics* **32**, 261–75.

Drechsler, M., Smith, G.H., Sturm, A., Wätzold, F., 2016. Cost-effectiveness of conservation payment schemes for species with different range sizes. *Conservation Biology* **30**, 894–9.

Drechsler, M., Wätzold, F., 2009. Applying tradable permits to biodiversity conservation: Effects of space-dependent conservation benefits and cost heterogeneity on habitat allocation. *Ecological Economics* **68**, 1083–92.

Drechsler, M., Wissel, C., 1997. Separability of local and regional dynamics in metapopulations. *Theoretical Population Biology* **51**, 9–21.

Drechsler, M., Zwerger, W., 1992. Crossover from BCS-superconductivity to Bose-condensation. *Annalen der Physik* **1**, 15–23.

Edwards-Jones, G., Davies, B., Hussain, S.S., 2000. *Ecological Economics: An Introduction.* Wiley-Blackwell.

Eeckhoudt, L., Schlesinger, H., Gollier, C., 2005. *Economic and Financial Decisions Under Risk.* Princeton University Press.

Egger, P., Nigai, S., 2015. Energy demand and trade in general equilibrium. *Environmental and Resource Economics* **60**, 191–213.

Eichhorn, M., Johst, K., Seppelt, R., Drechsler, M., 2012. Model-based estimation of collision risks of predatory birds with wind turbines. *Ecology and Society* **17** (2), 1.

Eisenack, K., 2016. Institutional adaptation to cooling water scarcity for thermo-electric power generation under global warming. *Ecological Economics* **124**, 153–63.

Eken, G., Bennun, L., Brooks, T.M., et al., 2004. Key biodiversity areas as site conservation targets. *BioScience* **54**, 1110‑18.

Engel, S., Pagiola, S., Wunder, S., 2008. Designing payments for environmental services in theory and practice: An overview of the issues. *Ecological Economics* **65**, 663–74.

Epanchin-Niell, R.S., Wilen, J.E., 2015. Individual and cooperative management of invasive species in human-mediated landscapes. *American Journal of Agricultural Economics* **97**, 180–98.

Epps, T.W., 2015. *Probability and Statistical Theory for Applied Researchers.* World Scientific Publishing Company.

Erev, I., Roth, A.E., 1998. Predicting how people play games with unique, mixed strategy equilibria. *American Economic Review* **88**, 848–81.

Faber, M., 2008. How to be an Ecological Economist. *Ecological Economics* **66**, 1–7.

Fahse, L., Wissel, C., Grimm, V., 1998. Reconciling classical and individual-based approaches in theoretical population ecology: A protocol for extracting population parameters from individual-based models. *American Naturalist* **152**, 838–52.

Fees, E., Seliger, A., 2013. *Umweltökonomie und Umweltpolitik.* Franz Vahlen, Munich, 4th ed.

Fehr, E., Schmidt, K.M., 1999. A theory of fairness, competition, and cooperation. *Quarterly Journal of Economics* **114**, 817–68.

Fenichel, E.P., Horan, R.D., 2016. Tinbergen and tipping points: Could some thresholds be policy-induced? *Journal of Economic Behavior and Organization* **132**, 137–52.

Fernandes, L., Day, J., Lewis, A., et al., 2005. Establishing representative no-take areas in the Great Barrier Reef: Large-scale implementation of theory on marine protected areas. *Conservation Biology* **19**, 1733–44.

Ferraro, P.J., Simpson, R.D., 2002. The cost-effectiveness of conservation payments. *Land Economics* **78**, 339–53.

Ferson, S., Burgman, M.A., 1995. Correlations, dependency bounds and extinction risks. *Biological Conservation* **73**, 101–5.

Fieberg, J., Ellner, S.P., 2001. Stochastic matrix models for conservation and management: A comparative review of methods. *Ecology Letters* **4**, 244–66.

Filatova, T., Verburg, P.H., Parker, D.C., Stannard, C.A., 2013. Spatial agent-based models for socio-ecological systems: Challenges and prospects. *Environmental Modelling & Software* **45**, 1–7.

Fortuna, M.A, Gómez-Rodríguez, C., Bascompte, J., 2006. Spatial network structure and amphibian persistence in stochastic environments. *Proceedings of the Royal Society B* **273**, 1429–34.

Fox, J., Nino-Murcia, A., 2005. Status of species conservation banking in the United States. *Conservation Biology* **19**, 996–1007.

Frank, K., 2005. Metapopulation persistence in heterogeneous landscapes: Lessons about the effect of stochasticity. *American Naturalist* **165**, 274–388.

Frederick, S., Loewenstein, G., O'Donoghue, T., 2002. Time discounting and time preference: A critical review. *Journal of Economic Literature* **40**, 351–401.

Fudenberg, D., Tirole, J., 1991. *Game Theory*. MIT Press.

Fung, T., O'Dwyer, J.P., Chisholm, R.A., 2018. Quantifying species extinction risk under temporal environmental variance. *Ecological Complexity* **34**, 139–46.

Gahvari, F., 2014. Second-best Pigouvian taxation: A clarification. *Environmental and Resource Economics* **59**, 525–35.

Gallai, N., Salles, J.-M., Settele, J., Vaissière, B.E., 2009. Economic valuation of the vulnerability of world agriculture confronted with pollinator decline. *Ecological Economics* **68**, 810–21.

Gardner, M., 1970. The fantastic combinations of John Conway's new solitaire game 'Life'. *Scientific American* **223**, 120–3.

Ge, X., Li, Y., Luloff, A., Dong, K., Xiao, J., 2015. Effect of agricultural economic growth on sandy desertification in Horqin Sandy Land. *Ecological Economics* **119**, 53–63.

Gerigk, J., MacKenzie, I.A., Ohndorf, M., 2015. A model of benchmarking regulation: Revisiting the efficiency of environmental standards. *Environmental and Resource Economics* **62**, 59–82.

Geritz, S.A.H., Kisdi, É., 2004. On the mechanistic underpinning of discrete-time population models with complex dynamics. *Journal of Theoretical Biology* **228**, 261–9.

Getz, W.M., 1998. An introspection on the art of modeling in population ecology. *BioScience* **48**, 540–52.

Getzin, S., Yizhaq, H., Bell, B., et al., 2016. Discovery of fairy circles in Australia supports self-organization theory. *Proceedings of the National Academy of Sciences of the USA* **113**, 3551–6.

Gintis, H., 2009. *Game Theory Evolving: A Problem-Centered Introduction to Modeling Strategic Interaction*. Princeton University Press, 2nd ed.

Gjertsen, H., Squires, D., Dutton, P.H., Eguchi, T., 2014. Cost-effectiveness of alternative conservation strategies with application to the Pacific leatherback turtle. *Conservation Biology* **28**, 140–9.

Glynatsi, N.E., Knight, V., Lee, T.E., 2018. An evolutionary game theoretic model of rhino horn devaluation. *Ecological Modelling* **389**, 33–40.

Goel, N.S., Richter-Dyn, N., 1974. *Stochastic Models in Biology.* Academic Press.

Gourguet, S., Macher, C., Doyen, L., et al., 2013. Managing mixed fisheries for bio-economic viability. *Fisheries Research* **140**, 46–62.

Grant, T.J., Parry, H.R., Zalucki, M.P., Bradbury, S.P., 2018. Predicting monarch butterfly (*Danaus plexippus*) movement and egg-laying with a spatially-explicit agent-based model: The role of monarch perceptual range and spatial memory. *Ecological Modelling* **374**, 37–50.

Grashof-Bokdam, C.J., Cormont, A., Polman, N.B.P., Westerhof, E.J.G.M., Franke, J.G.J., Opdam, P.F.M., 2017. Modelling shifts between mono- and multifunctional farming systems: The importance of social and economic drivers. *Landscape Ecology* **32**, 595–607.

Grimm, V., 1999. Ten years of individual-based modelling in ecology: What have we learned and what could we learn in the future? *Ecoogical Modelling* **115**, 129–48.

Grimm, V., Berger, U., 2016. Robustness analysis: Deconstructing computational models for ecological theory and applications. *Ecological Modelling* **326**, 162–7.

Grimm, V., Berger, U., Bastiansen, F., et al., 2006. A standard protocol for describing individual-based and agent-based models. *Ecological Modelling* **198**, 115–26.

Grimm, V., Railsback, S.F., 2005. *Individual-Based Modelling and Ecology.* Princeton University Press.

Grimm, V. Revilla. E., Berger, Jeltsch, et al., 2005. Pattern-oriented modelling of agent-based complex systems: Lessons from ecology. *Science* **310**, 987–91.

Grimm, V., Wissel, C., 1997. Babel, or the ecological stability discussions: An inventory and analysis of terminology and a guide for avoiding confusion. *Oecologia* **109**, 323–34.

Grimm, V. Wissel, V, 2004. The intrinsic mean time to extinction: A unifying approach to analysing persistence and viability of populations. *Oikos* **105**, 501–11.

Groeneveld, J., Müller, B., Buchmann, C.M., et al., 2017. Theoretical foundations of human decision-making in agent-based land use models: A review. *Environmental Modelling & Software* **87**, 39–48.

Grovermann, C., Schreinemachers, P., Riwthong, S., Berger, T., 2017. 'Smart' policies to reduce pesticide use and avoid income trade-offs: An agent-based model applied to Thai agriculture. *Ecological Economics* **132**, 91–103.

Gunton, R.M., Marsh, C.J., Moulherat, S., et al., 2017. Multicriterion trade-offs and synergies for spatial conservation planning. *Journal of Applied Ecology* **54**, 903–13.

Güth, W., Schmittberger, R., Schwarze, B., 1982. An experimental analysis of ultimatum bargaining. *Journal of Economic Behavior & Organization* **3**, 367–88.

Gutrich, J.J., Gigliello, K., Gardner, K.V., Elmore, A.J., 2016. Economic returns of groundwater management sustaining an ecosystem service of dust suppression by alkali meadow in Owens Valley, California. *Ecological Economics* **121**, 1–11.

Haab, T.C., Whitehead, J.C., 2014. *Environmental and Natural Resource Economics: An Encyclopedia*. Greenwood Publishing Group, 1st ed.

Hackett, S.B., Moxnes, E., 2015. Natural capital in integrated assessment models of climate change. *Ecological Economics* **116**, 354–61.

Hailu, A., Schilizzi, S., Thoyer, S., 2005. Assessing the performance of auctions for the allocation of conservation contracts: Theoretical and computational approaches. Selected paper prepared for presentation at the American Agricultural Economics Association Annual Meeting, Providence, Rhode Island, 24–27 July 2005. www.researchgate.net/profile/Steven_Schilizzi/publication/23506218_Assessing_the_performance_of_auctions_for_the_allocation_of_conservation_contracts_Theoretical_and_computational_approaches/links/0fcfd50ad593cb5a61000000/Assessing-the-performance-of-auctions-for-the-allocation-of-conservation-contracts-Theoretical-and-computational-approaches.pdf (last accessed 10 October 2019).

Haken, H., 1991. *Synergetics*. Springer.

Hamill, L., Gilbert, N., 2016. *Agent-Based Modelling in Economics*. Wiley.

Hanau, A., 1928. *Die Prognose der Schweinepreise. Vierteljahreshefte zur Konjunkturforschung*. Verlag Reimar Hobbing.

Hanley, N., Acs, S., Dallimer, M., Gaston, K.J., Graves, A., Morris, J., Armsworth, P.R., 2012. Farm-scale ecological and economic impacts of agricultural change in the uplands. *Land Use Policy* **29**, 587–97.

Hanley, N., Shogren, J.F., White, B., 2007. *Envoronmental Economics: In Theory and Practice*. Palgrave Macmillan, 2nd ed.

Hansjürgens, B., Antes, R., Strunz, M., (eds.), 2016. *Permit Trading in Different Applications*. Routledge.

Hanski, I., 1994. A practical model of metapopulation dynamics. *Journal of Animal Ecology* **63**, 151–62.

Hanski, I., 1999. *Metapopulation Ecology*. Oxford University Press.

Hanski, I., Gaggiotti, O.E., (eds.), 2004. *Ecology, Genetics and Evolution of Metapopulations*. Elsevier Academic Press.

Hanski, I., Gilpin, M. (eds.), 1997. *Metapopulation Biology*. Academic Press.

Hanski, I., Ovaskainen, O., 2000. The metapopulation capacity of a fragmented landscape. *Nature* **404**, 755–8.

Hartig, F., Calabrese, J.M., Reineking, B., 2011. Statistical inference for stochastic simulation models: Theory and application. *Ecology Letters* **14**, 816–27.

Hartig, F., Drechsler, M., 2009. Smart spatial incentives for market-based conservation. *Biological Conservation* **142**, 779–88.

Hartig, F., Drechsler, M., 2010. Stay by thy neighbor? Social organization determines the efficiency of biodiversity markets with spatial incentives. *Ecological Complexity* **7**, 91–9.

Hartig, F., Horn, M., Drechsler, M., 2010. EcoTRADE: A multi-player network game of a tradable permit market for biodiversity credits. *Environmental Modelling & Software* **25**, 1479–80.

Hassell, M.P., 1975. Density-dependence in single-species populations. *Journal of Animal Ecology* **44**, 283–95.

Hatcher, A., Nøstbakken, L., 2015. Quota setting and enforcement choice in a shared fishery. *Environmental and Resource Economics* **61**, 559–75.

Heckbert, S., Baynes, T., Reeson, A., 2010. Agent-based modeling in ecological economics. *Annals of the New York Academy of Sciences* **1185**, 39–53.

Heinrichs, J.A., Aldridge, C.L., O'Donnell, M.S., Schumaker, N.H., 2017. Using dynamic population simulations to extend resource selection analyses and prioritize habitats for conservation. *Ecological Modelling* **359**, 449–59.

Hicks, J.S., Burgman, M.A., Marewski, J.N., Fidler, F., Gigerenzer, G., 2012. Decision making in a human population living sustainably. *Conservation Biology* **26**, 760–8.

Hoffman, E., McCabe, K., Smith, V.L., 1996. Social distance and other-regarding behavior in dictator games. *American Economic Review* **86**, 653–60.

Hohenberg, P.C., Krekhov, H.P., 2015. An introduction to the Ginzburg–Landau theory of phase transitions and nonequilibrium patterns. arXiv:1410.7285v3 [cond-mat.stat-mech] 18 February 2015.

Holland, M.D., Hastings, A., 2008. Strong effect of dispersal network structure on ecological dynamics. *Nature* **456**, 792–4.

Holling, C.S., 1973. Resilience and stability of ecological systems. *Annual Review of Ecology and Systematics* **4**, 1–23.

Holyoak, M., Leibold, M.A., Holt, R.D. (eds.), 2005. *Metacommunities: Spatial Dynamics and Ecological Communities*. University of Chicago Press.

Horn, J., Becher, M. A., Kennedy, P.J., Osborne, J.L., Grimm, V., 2016. Multiple stressors: Using the honeybee model BEEHAVE to explore how spatial and temporal forage stress affects colony resilience. *Oikos* **125**, 1001–16.

Huang, B., Abbott, J.K., Fenichel, E.P., et al., 2017. Testing the feasibility of a hypothetical whaling-conservation permit market in Norway. *Conservation Biology* **31**, 809–17.

Hubbell, S.P., 2001. *The Unified Neutral Theory of Biodiversity and Biogeography*. Princeton University Press.

Iacona, G.D., Bode, M., Armsworth, P.R., 2016. Limitations of outsourcing on-the-ground biodiversity conservation. *Conservation Biology* **30**, 1245–54.

Iftekhar, Md.S., Latacz-Lohmann, U., 2017. How well do conservation auctions perform in achieving landscape-level outcomes? A comparison of auction formats and bid selection criteria. *Australian Journal of Agricultural and Resource Economics* **61**, 557–75.

Iftekhar, Md.S., Tisdell, J.G., 2016. An agent based analysis of combinatorial bidding for spatially targeted multi-objective environmental programs. *Environmental and Resource Economics* **64**, 537–58.

Iho, A., Ahlvik, L., Ekholm, P., Lehtoranta, J., Kortelainen, P., 2017. Optimal phosphorus abatement redefined: Insights from coupled element cycles. *Ecological Economics* **137**, 13–19.

Ings, T.C., Montoya, J.M., Bascompte, J., et al., 2009. Ecological networks: Beyond food webs. *Journal of Animal Ecology* **79**, 253–69.

Innes, R., 2000. The economics of takings and compensation when land and its public use values are in private hands. *Land Economics* **76**, 195–212.

Ishizaka, A., Nemery, P., 2013. *Multi-Criteria Decision Analysis: Methods and Software*. Wiley.

Jackson, N.D., Fahrig, L., 2016. Habitat amount, not habitat configuration, best predicts population genetic structure in fragmented landscapes. *Landscape Ecology* **31**, 951–68.

Jalas, M., Juntunen, J.K., 2015. Energy intensive lifestyles: Time use, the activity patterns of consumers, and related energy demands in Finland. *Ecological Economics* **113**, 51–9.

Jeltsch, F., Milton, S.J., Dean, W.R.J., von Rooyen, N., 1996. Tree spacing and coexistence in semiarid savannas. *Journal of Ecology* **84**, 583–95.

Jeltsch, F., Weber, G.E., Grimm, V., 2000. Ecological buffering mechanisms in savannas: A unifying theory of long-term tree–grass coexistence. *Plant Ecology* **161**, 161–71.

Jiang, Y., Koo, W.W., 2014. The short-term impact of a domestic cap-and-trade climate policy on local agriculture: A policy simulation with producer behavior. *Environmental and Resource Economics* **58**, 511–37.

Johnson, R.A., Wichern, D.W., 2013. *Applied Multivariate Statistical Analysis: Pearson New International Edition*. Pearson Education Limited, 6th ed.

Johst, K., Drechsler, M., Thomas, J., Settele, J., 2006. Influence of mowing on the persistence of two endangered large blue butterfly species. *Journal of Applied Ecology* **43**, 333–42.

Jopp, F., Reuter, F., Breckling, B. (eds.), 2011. *Modelling Complex Ecological Dynamics*. Springer.

Kælbling, L.P., Littman, M.L., Moore, A.W., 1996. Reinforcement learning: A survey. *Journal of Artificial Intelligence Research* **4**, 237–85.

Kakeu, J., Johnson, E.P., 2018. Information exchange and transnational environmental problems. *Environmental and Resource Economics* **71**, 583–604.

Kalkuhl, M., Edenhofer, M., Lessmann, K., 2015. The role of carbon capture and sequestration policies for climate change mitigation. *Environmental and Resource Economics* **60**, 55–80.

Kane, A., Jackson, A.L., Ogada, D.L., Monadjem, A., McNally, L., 2014. Vultures acquire information on carcass location from scavenging eagles. *Proceedings of the Royal Society B: Biological Sciences* **281** (1793), 20141072.

Kirkpatrick, S., Gelatt, C.D., Vecchi, M.P., 1983. Optimization by simulated annealing. *Science* **220**, 671–80.

Knight, F., 1921. *Risk, Uncertainty, and Profit*. Houghton Mifflin.

Koh, I., Lonsdorf, E.V., Artz, D.R., Pitts-Singer, T.L., Ricketts, T.H., 2018. Ecology and economics of using native managed bees for almond pollination. *Journal of Economic Entomology* **111**, 16–25.

Köhler, P., Huth, A., 1998. The effects of tree species grouping in tropical rainforest modelling: Simulations with the individual-based model FORMIND. *Ecological Modelling* **109**, 301–21.

Kollock, P., 1998. Social dilemmas: The anatomy of cooperation. *Annual Review of Sociology* **24**, 183–214.

Kontoleon, A., Pascual, U., Swanson, T. (eds.), 2007. *Biodiversity Economics: Principles, Methods, Applications*. Cambridge University Press.

Kragt, M.E., Robertson, M.J., 2014. Quantifying ecosystem services trade-offs from agricultural practices. *Ecological Economics* **102**, 147–57.

Krämer, J.E., Wätzold, F., 2018. The agglomeration bonus in practice: An exploratory assessment of the Swiss network bonus. *Journal for Nature Conservation* **43**, 126–35.

Kuhfuss, L., Préget, R., de Vries, F.P., Hanley, N., Thoyer, S., 2018. Nudging participation and spatial agglomeration in payment for environmental service schemes. Discussion Papers in Environment and Development Economics, School of Geography and Sustainable Development, University of St Andrews. www.st-andrews.ac.uk/media/dept-of-geography-and-sustainable-development/pdf-s/DP%202017-11%20Kuhfuss%20et%20alpdf.pdf (last accessed 10 October 2019).

Kuhfuss, L., Préget, R., Thoyer, S., Hanley, N., 2016. Nudging farmers to enrol land into agri-environmental schemes: The role of a collective bonus. *European Review of Agricultural Economics* **43**, 609–36.

Kukkala, A.S., Arponen, A., Maiorano, L., et al., 2016. Matches and mismatches between national and EU-wide priorities: Examining the Natura 2000 network in vertebrate species conservation. *Biological Conservation* **198**, 193–201.

Laffont, J.-J., Martimort, D., 2002. *The Theory of Incentives: The Principal–Agent Model*. Princeton University Press.

Laffont, J.-J., Tirole, J., 1993. *A Theory of Incentives in Procurement and Regulation*. MIT Press.

Lapola, D.M., Schaldach, R., Alcamo, J., et al., 2010. Indirect land-use changes can overcome carbon savings from biofuels in Brazil. *Proceedings of the National Academy of Sciences of the USA* **107**, 3388–93.

Laufenberg, J.S., Clark, J.D., Hooker, M.J., et al., 2016. Demographic rates and population viability of black bears in Louisiana. *Wildlife Monographs* **194**, 1–37.

Latacz-Lohmann, U., van der Hamsvoort, C., 1997. Auctioning conservation contracts: A theoretical analysis and an application. *American Journal of Agricultural Economics* **79**, 407–18.

Lehmann, P., Söderholm, P., 2018. Can technology-specific deployment policies be cost-effective? The case of renewable energy support schemes. *Environmental and Resource Economics* **71**, 475–505.

Lehtomäki, J., Moilanen, A., 2013. Methods and workflow for spatial conservation prioritization using Zonation. *Environmental Modelling & Software* **47**, 128–37.

Leibold, M.A., Chase, J.M., 2017. *Metacommunity Ecology*. Princeton University Press.

Leibold, M.A., Holyoak, M., Mouquet, N., et al., 2004. The metacommunity concept: A framework for multi-scale community ecology. *Ecology Letters* **7**, 601–13.

Lennox, G.D., Gaston, K.J., Acs, S., et al., 2013. Conservation when landowners have bargaining power: Continuous conservation investments and cost uncertainty. *Ecological Economics* **93**, 69–78.

Lentini, P.E., Gibbons, P., Carwardine, J., et al., 2013. Effect of planning for connectivity on linear reserve networks. *Conservation Biology* **4**, 796–807.

Levin, S.A., Paine, R.T., 1974. Disturbance, patch formation, and community structure. *Proceedings of the National Academy of Sciences of the USA* **71**, 2744–7.

Levins, R., 1966. The strategy of model building in population biology. *American Naturalist* **54**, 421–31.

Levins, R., 1969. Some demographic and genetic consequences of environmental heterogeneity for biological control. *Bulletin of the Entomological Society of America* **15**, 237–40.

Lewis, D.J., Plantinga, A.J., 2007. Policies for habitat fragmentation: Combining econometrics with GIS-based landscape simulations. *Land Economics* **83**, 109–27.

Lewis, D.J., Plantinga, A.J., Nelson, E., Polasky, S., 2011. The efficiency of voluntary incentive policies for preventing biodiversity loss. *Resource and Energy Economics* **33**, 192–211.

Leydold, J., 1997. Ein Cobweb-Model. statistik.wu-wien.ac.at/~leydold/MOK/HTML/node193.html (last accessed 10 October 2019).

Lifshitz, E.M., Pitaevskii, L.P., 1980. *Statistical Physics Part 2.* Butterworth-Heinemann.

Lin, J., Li, X., 2016. Conflict resolution in the zoning of eco-protected areas in fast-growing regions based on game theory. *Journal of Environmental Management* **170**, 177–85.

Liu, Z., Gong, Y., Kontoleon, A., 2018. How do payments for environmental services affect land tenure? Theory and evidence from China. *Ecological Economics* **144**, 195–213.

Lonsdorf, E.V., Thogmartin, W.E., Jacobi, S., et al., 2016. A generalizable energetics-based model of avian migration to facilitate continental-scale waterbird conservation. *Ecological Applications* **26**, 1136–53.

Lotka, A.J., 1920. Analytical note on certain rhythmic relations in organic systems. *Proceedings of the National Academy of Sciences of the USA* **6**, 410–15.

Louviere, J.J., Hensher, D.A., Swait, J.D., 2000. *Stated Choice Methods: Analysis and Application.* Cambridge University Press.

McDonald-Madden, E., Bode, M., Game, E.T., Grantham, H., Possingham, H.P., 2008. The need for speed: Informed land acquisitions for conservation in a dynamic property market. *Ecology Letters* **11**, 1169–77.

McElderry, R.M., Salvato, M.H., Horvitz, C.C., 2015. Population viability models for an endangered endemic subtropical butterfly: Effects of density and fire on population dynamics and risk of extinction. *Biodiversity and Conservation* **24**, 1589–608.

McGowan, C.P., Allan, N., Servoss, J., Hedwall, S., Wooldridge, B., 2017. Incorporating population viability models into species status assessment and listing decisions under the U.S. Endangered Species Act. *Global Ecology and Conservation* **12**, 119–30.

Mallard, G., 2016. *Bounded Rationality and Behavioural Economics.* Routledge.

Managi, S. (ed.), 2012. *The Economics of Biodiversity and Ecosystem Services.* Routledge, 1st ed.

Mangham, L.J., Hanson, K., McPake, B., 2009. How to do (or not to do) ... Designing a discrete choice experiment for application in a low-income country. *Health Policy and Planning* **24**, 151–8.

Marescot, L., Chapron, G., Chadès, I., et al., 2013. Complex decisions made simple: A primer on stochastic dynamic programming. *Methods in Ecology and Evolution* **4**, 872–84.

Margules, C.R., Pressey, R.L., 2000. Systematic conservation planning. *Nature* **405**, 243–53.

Martinez-Allier, J., Muradian, R. (eds.), 2015. *Handbook of Ecological Economics.* Edward Elgar.

Marzloff, M.P., Little, L.R., Johnson, C.R., 2016. Building resilience against climate-driven shifts in a temperate reef system: Staying away from context-dependent ecological thresholds. *Ecosystems* **19**, 1–15.

Matthews, R.B., Gilbert, N.G., Roach, A., Polhill, J.G., Gotts, N.M., 2007. Agent-based land-use models: A review of applications. *Landscape Ecology* **22**, 1447–59.

May, F., Huth, A., Wiegand, T., 2015. Moving beyond abundance distributions: Neutral theory and spatial patterns in a tropical forest. *Proceedings of the Royal Society B* **282**, 20141657.

May, R.M., 1976. Simple mathematical models with very complicated dynamics. *Nature* **261**, 459–67.

Maynard Smith, J., 1982. *Evolution and the Theory of Games.* Cambridge University Press.

Meadows, D.H., Meadows, D.L., Randers, J., Behrens, W.W. III., 1972. *The Limits to Growth: A Report for the Club of Rome's Project on the Predicament of Mankind.* Universe Books.

Metcalfe, K., Vaz, S., Engelhard, G.H., Villanueva, M.C., Smith, R.J., Mackinson, S., 2015. Evaluating conservation and fisheries management strategies by linking spatial prioritization software and ecosystem and fisheries modelling tools. *Journal of Applied Ecology* **52**, 665–74.

M'Gonigle, L.K., Greenspoon, P.B., 2014. Allee effects and species co-existence in an environment where resource abundance varies. *Journal of Theoretical Biology* **361**, 61–8.

Millennium Ecosystem Assessment, 2005. Ecosystems and human well-being: Biodiversity synthesis. World Resources Institute.

Mitchell, R.C., Carson, R.T., 1989. *Using Surveys to Value Public Goods: The Contingent Valuation Method.* Resources for the Future, Washington, DC.

Miyasaka, T., Le, Q.B., Okuro, T., Zhao, X., Takeuchi, K., 2017. Agent-based modeling of complex social–ecological feedback loops to assess multi-dimensional trade-offs in dryland ecosystem services. *Landscape Ecology* **32**, 707–27.

Moilanen, A., Wilson, K.A., Possingham, H.P., 2009. *Spatial Conservation Prioritization.* Oxford University Press.

Moloney, K.A., Levin, S.A., 1996. The effects of disturbance architecture on landscape-level population dynamics. *Ecology* **77**, 375–94.

Morrison, M., Morgan, M.S., 1999a. Introduction. In Morgan, M.S., Morrison, M. (eds.), *Models as Mediators.* Cambridge University Press, pp. 1–9.

Morrison, M., Morrison, M.S., 1999b. Models as mediating instruments. In Morgan, M.S., Morrison, M. (eds.), *Models as Mediators.* Cambridge University Press, pp. 10–37.

Mouysset, L., Doyen, L., Allaire, J.G., Leger, F., 2011. Bio economic modeling for a sustainable management of biodiversity in agricultural lands. *Ecological Economics* **70**, 617–26.

Mouysset, L., Doyen, L., Jiguet, F., 2013. From population viability analysis to coviability of farmland biodiversity and agriculture. *Conservation Biology* **28**, 187–201.

Moxey, A., White, B., Ozanne, A., 1999. Efficient contract design for agri-environment policy. *Journal of Agricultural Economics* **50**, 187–202.

Naidoo, R., Balmford, A., Ferraro, P.J., et al.., 2006. Integrating economic cost into conservation planning. *Trends in Ecology and Evolution* **21**, 681–7.

Nalle, D.J., Montgomery, C.A., Arthur, J.L., Polasky, S., Schumaker, N.H., 2004. Modeling joint production of wildlife and timber. *Journal of Environmental Economics and Management* **48**, 997–1017.

Negele, J.W., Orland, H., 1988. *Quantum Many-Particle Systems*. Perseus.

Nelson, E., Polasky, S., Lewis, D.J., et al., 2008. Efficiency of incentives to jointly increase carbon sequestration and species conservation on a landscape. *Proceedings of the National Academy of Sciences of the USA* **105**, 9471–6.

Ng, C.F., McCarthy, M.A., Martin, T.G., Possingham, H.P., 2014. Determining when to change course in management actions. *Conservation Biology* **6**, 1617–25.

Nguyen, N.P., Shortle, J.S., Reed, P.M., Nguyen, T.T., 2013. Water quality trading with asymmetric information, uncertainty and transaction costs: A stochastic agent-based simulation. *Environmental and Resource Economics* **35**, 60–90.

Nicholson, C.R., Starfield, A.M., Kofinas, G.P., Kruse, J.A., 2002. Ten heuristics for interdisciplinary modeling projects. *Ecosystems* **5**, 376–84.

Ninan, K.N. (ed.), 2010. *Conserving and Valuing Ecosystem Services and Biodiversity: Economic, Institutional and Social Challenges*. Routledge, 1st ed.

Nisbet, R.M., Gurney, W., 2003. *Modelling Fluctuating Populations*. Blackburn, reprint of 1st ed. (1982).

Nowak, M., 2006. Five rules for the evolution of cooperation. *Science* **314**, 1560–3.

Nowak, M.A., May, R.M., 1992. Evolutionary games and spatial chaos. *Nature* **246**, 15–18.

Nunes, P., van den Bergh, J.C.J.M., Nijkamp, P., 2003. *The Ecological Economics of Biodiversity: Methods and Policy Applications*. Edward Elgar.

Nuñez, T.A., Lawler, J.J., McRae, B.H., et al., 2013. Connectivity planning to address climate change. *Conservation Biology* **27**, 407–16.

O'Hara, R.B., Arjas, E., Toivonen, H., Hanski, I., 2002. Bayesian analysis of metapopulation data. *Ecology* **83**, 2408–15.

Ohl, C., Drechsler, M., Johst, K., Wätzold, F., 2006. Managing land use and land cover change in the biodiversity context with regard to efficiency, equality and ecological effectiveness. *UFZ-Diskussionspapiere* 3/2006, UFZ Leipzig-Halle GmbH, Germany. www.ufz.de/export/data/2/26205_dp200603.pdf (last accessed 10 October 2019).

Ohl, C., Drechsler, M., Johst, K., Wätzold, F., 2008. Compensation payments for habitat heterogeneity: Existence, efficiency, and fairness considerations. *Ecological Economics* **67**, 126–74.

Oleś, K., Gudowska-Nowak, E., Kleczkowski, A., 2013. Efficient control of epidemics spreading on networks: Balance between treatment and recovery. *PloS one* **8**, e63813.

Ozanne, A., Hogan, T., Colman, D., 2001. Moral hazard, risk aversion and compliance monitoring in agri-environmental policy. *European Review of Agricultural Economics* **28**, 329–47.

Parkhurst, G.M., Shogren, J.F., 2007. Spatial incentives to coordinate contiguous habitat. *Ecological Economics* **64**, 344–55.

Parkhurst, G.M., Shogren, J.F., Basrian, C., et al., 2002. Agglomeration bonus: An incentive mechanism to reunite fragmented habitat for biodiversity conservation. *Ecological Economics* **41**, 305–28.

Parkhurst, G.M., Shogren, J.F., Crocker, T., 2016. Tradable set-aside requirements (TSARs): Conserving spatially dependent environmental amenities. *Environmental and Resource Economics* **63**, 719–44.

Pascual, U., Muradian, R., Rodríguez, L. C., Duraiappah, A., 2010. Exploring the links between equity and efficiency in payments for environmental services: A conceptual approach. *Ecological Economics* **69**, 1237–44.

Pautrel, X., 2015. Abatement technology and the environment–growth nexus with education. *Environmental and Resource Economics* **61**, 297–318.

Pe'er, G., Saltz, D., Münkemüller, T., Matsinos, Y.G., Thulke, H.-H., 2013. Simple rules for complex landscapes: The case of hilltopping movements and topography. *Oikos* **122**, 1483–95.

Peretti, C.T., Munch, S.B., Sugihara, G., 2013. Model-free forecasting outperforms the correct mechanistic model for simulated and experimental data. *Proceedings of the National Academy of Sciences of the USA* **110**, 5253–7.

Perrings, C., 2006. Resilience and sustainable development. *Environmental and Development Economics* **11**, 417–27.

Pezzey, J.C.V., Burke, P.J., 2014. Towards a more inclusive and precautionary indicator of global sustainability. *Ecological Economics* **106**, 141–54.

Pielou, E.C., 1984. *The Interpretation of Ecological Data: A Primer on Classification and Ordination.* Wiley.

Piñero, P., Heikkinen, M., Mäenpää, I., Pongrácz, E., 2015. Sector aggregation bias in environmentally extended input output modeling of raw material flows in Finland. *Ecological Economics* **119**, 217–29.

Plumecocq, G., 2014. The second generation of ecological economics: How far has the apple fallen from the tree? *Ecological Economics* **107**, 457–68.

Polasky, S., Lewis, D.J., Plantinga, A.J., Nelson, E., 2014. Implementing the optimal provision of ecosystem services. *Proceedings of the National Academy of Sciences of the USA* **111**, 6248–53.

Polasky, S., Nelson, E., Camm, J., et al., 2008. Where to put things? Spatial land management to sustain biodiversity and economic returns. *Biological Conservation* **141**, 1505–24.

Polasky, S., Segerson, K., 2009. Integrating ecology and economics in the study of ecosystem services: Some lessons learned. *Annual Review of Resource Economics* **1**, 409–34.

Přibylová, L., 2018. Regime shifts caused by adaptive dynamics in prey-predator models and their relationship with intraspecific competition. *Ecological Complexity* **36**, 48–56.

Proctor, W., Drechsler, M., 2006. Deliberative multicriteria evaluation. *Environmental Planning C Government and Policy* **24**, 169–90.

Proops, J.L.R., 1989. Ecological economics: Rationale and problem areas. *Ecologcial Economics* **1**, 59–76.

Proops, J., Safonov, P. (eds.), 2004. *Modelling in Ecological Economics.* Edward Elgar.

Quaas, M.F., Baumgärtner, S., Becker, C., Frank, K., Müllser, B., 2007. Uncertainty and sustainability in the management of rangelands. *Ecological Economics* **62**, 251–66.

Quaas, M.F., van Soost, D., Baumgärtner, S., 2013. Complementarity, impatience, and the resilience of natural-resource-dependent economies. *Journal of Environmental Economics and Management* **66**, 15–32.

Radchuk, V., Oppel, S., Groeneveld, J., Grimm, V., Schtickzelle, N., 2016. Simple or complex: Relative impact of data availability and model purpose on the choice of model types for population viability analyses. *Ecological Modelling* **323**, 87–95.

Rademacher, C., Neuert, C., Grundmann, V., Wissel, C., Grimm, V., 2004. Reconstructing spatiotemporal dynamics of Central European natural beech forests: The rule-based forest model BEFORE. *Forest Ecology and Management* **194**, 349–68.

Raffin, N., Seegmuller, T., 2017. The cost of pollution on longevity, welfare and economic stability. *Environmental and Resource Economics* **68**, 683–704.

Railsback, S.F., Grimm, V., 2012. *Agent-Based and Individual-Based Modeling: A Practical Introduction.* Princeton University Press.

Railsback, S.F., Harvey, B.C., 2002. Analysis of habitat selection rules using an individual-based model. *Ecology* **83**, 1817–30.

Rämö, J., Tahvonen, O., 2017. Optimizing the harvest timing in continuous cover forestry. *Environmental and Resource Economics* **67**, 853–68.

Reeson, A., Capon, T., Whitten, S., 2018. Are auctions for environmental service provision likely to suffer with repetition? Paper presented at the World Conference of Environmental and Resource Economists (WCERE), June 2018, Gothenburg. fleximeets.com/wcere2018/?p=programme (last accessed 10 October 2019).

Reeson, A.F., Rodriguez, L.C., Whitten, S.M., et al., 2011. Adapting auctions for the provision of ecosystem services at the landscape scale. *Ecological Economics* **70**, 1621–7.

Regan, H.M., Bohórquez, C.I., Keith, D.A., Regan, T.J., Anderson, K.E., 2017. Implications of different population model structures for management of threatened plants. *Conservation Biology* **31**, 459–68.

Reznik, A., Feinerman, E., Finkelshtain, I., et al., 2017. Economic implications of agricultural reuse of treated wastewater in Israel: A statewide long-term perspective. *Ecological Economics* **135**, 222–33.

Ring, I., Drechsler, M., van Teeffelen, A.J.A., Irawan, S., Venter, O., 2010. Biodiversity conservation and climate mitigation: What role can economic instruments play? *Current Opinion in Environmental Sustainability* **2**, 50–8.

Rinnan, D.S., 2018. Population persistence in the face of climate change and competition: A battle on two fronts. *Ecological Modelling* **385**, 78–88.

Rivers, N., Growes, S., 2013. The welfare impact of self-supplied water pricing in Canada: A computable general equilibrium assessment. *Environmental and Resource Economics* **55**, 419–45.

Robert, A., 2009. The effects of spatially correlated perturbations and habitat configuration on metapopulation persistence. *Oikos* **118**, 1590–600.

Robinson, K.F., Fuller, A.K., Schiavone, M.V., et al., 2017. Addressing wild turkey population declines using structured decision making. *Journal of Wildlife Management* **81**, 393–405.

Rode, J., Wittmer, H., Emerton, L., Schröter-Schlaack, C., 2016. 'Ecosystem service opportunities': A practice-oriented framework for identifying economic instruments to enhance biodiversity and human livelihoods. *Journal for Nature Conservation* **33**, 35–47.

Rodrigues, A.S.L., Akçakaya, H.R., Andelman, S.J., et al., 2004. Global gap analysis: Priority regions for expanding the gobal protected-area network. *BioScience* **54**, 1092–100.

Rodrigues, A.S.L., Pilgrim, J.D., Lamoreux, J.F., Hoffmann, M., Brooks, T.M., 2006. The value of the IUCN Red List for conservation. *Trends in Ecology & Evolution* **21**, 71–6.

Rodríguez, J.P., Rodríguez-Clark, K.M., Baillie, J.E.M., et al., 2010. Establishing IUCN Red List criteria for threatened ecoystems. *Conservation Biology* **25**, 21–9.

Roman, S., Bulock, S., Brede, M., 2017. Coupled societies are more robust against collapse: A hypothetical look at Easter Island. *Ecological Economics* **132**, 264–78.

Røpke, I., 2005. Trends in the development of ecological economics from the late 1980s to the early 2000s. *Ecologcial Economics* **55**, 262–90.

Rosindell, J., Hubbell, S.P., Etienne, R.S., 2011. The Unified Neutral Theory of Biodiversity and Biogeography at age ten. *Trends in Ecology and Evolution* **26**, 340–8.

Roth, A.E., Erev, I., 1995. Learning in extensive form games: Experimental data and simple dynamics models in the intermediate term. *Games and Economic Behaviour* **8**, 164–212.

Rout, T.M., Baker, C.M., Huxtable, S., Wintle, B.A., 2017. Monitoring, imperfect detection, and risk optimization of a Tasmanian devil insurance population. *Conservation Biology* **32**, 267–75.

Rubinstein, A., 1998. *Modelling Bounded Rationality*. MIT Press.

Rueda-Cediel, P., Anderson, K.E., Regan, T.J., Regan, H.M., 2018. Effects of uncertainty and variability on population declines and IUCN Red List classifications. *Conservation Biology* **32**, 916–25.

Ryberg, W.A., Hill, M.T., Painter, C.W., Fitzgerald, L.A., 2014. Linking irreplaceable landforms in a self-organizing landscape to sensitivity of population vital rates for an ecological specialist. *Conservation Biology* **29**, 888–98.

Rytteri, S., Kuussaari, M., Saastamoinen, M., Ovaskainen, O., 2017. Can we predict the expansion rate of a translocated butterfly population based on a priori estimated movement rates? *Biological Conservation* **215**, 189–95.

Saltelli, A., Chan, K., Scott, E.M., 2009. *Sensitivity Analysis*. Wiley.

Sanga, G.J., Mungatana, E.D., 2016. Integrating ecology and economics in understanding responses in securing land-use externalities internalization in water catchments. *Ecological Economics* **121**, 28–39.

dos Santos, F.A.S., Johst, K., Grimm, V., 2011. Neutral communities may lead to decreasing diversity disturbance relationships: Insights from a generic simulation model. *Ecology Letters* **14**, 653–60.

Santos, R., Schröter-Schlaack, C., Antunes, P., Ring, I., Clemente, P., 2014. Reviewing the role of habitat banking and tradable development rights in the conservation policy mix. *Environmental Conservation* **42**, 294–305.

Scheiter, S., Schulte, J., Pfeiffer, M., et al.., 2019. How does climate change influence the economic value of ecosystem services in savanna rangelands? *Ecological Economics* **157**, 342–56.

Schelling, T.C., 1969. Models of segregation. *American Economic Review* **59**, 488–93.

Schirpke, U., Marino, D., Marucci, A., Palmieri, M., 2018. Positive effects of payments for ecosystem services on biodiversity and socio-economic development: Examples from Natura 2000 sites in Italy. *Ecosystem Services* **34**, 96–105.

Schlüter, M., Baeza, A., Dressler, G., et al., 2017. A framework for mapping and comparing behavioural theories in models of social-ecological systems. *Ecological Economics* **131**, 21–35.

Schlüter, M., Pahl-Wostl., C., 2007. Mechanisms of reslience in common-pool resource management system: An agent-based model of water use in a river basin. *Ecology and Society* **12**(2), 4.

Schomers, S., Matzdorf, B., 2013. Payments for ecosystem services: A review and comparison of developing and industrialized countries. *Ecosystem Services* **6**, 16–30.

Schöttker, O., Johst, K., Drechsler, M., Wätzold, F., 2016. Land for biodiversity conservation: To buy or borrow? *Ecological Economics* **129**, 94–103.

Schulze, J., Frank, K., Müller, B., 2016. Governmental response to climate risk: Model-based assessment of livestock supplementation in drylands. *Land Use Policy* **54**, 47–57.

Schulze, J., Martin, R., Finger, A., et al., 2015. Design, implementation and test of a serious online game for exploring complex relationships of sustainable land management and human well-being. *Environmental Modelling & Software* **65**, 58–66.

Settle, C., Crocker, T.D., Shogren, J.F., 2002. On the joint determination of biological and economic systems. *Ecological Economics* **42**, 301–11.

Sharma, Y., Abbott, K.C., Dutta, P.S., Gupta, A.K., 2015. Stochasticity and bistability in insect outbreak dynamics. *Theoretical Ecology* **8**, 163–74.

Sheng, J., Qiu, H., 2018. Governmentality within REDD+: Optimizing incentives and efforts to reduce emissions from deforestation and degradation. *Land Use Policy* **76**, 611–22.

Sheng, J., Wu, Y., Zhang, M., Miao, Z., 2017. An evolutionary modeling approach for designing a contractual REDD+ payment scheme. *Ecological Indicators* **79**, 276–85.

Shmueli, G., 2010. To explain or to predict. *Statistical Science* **25**, 289–310.

Shoham, Y., Leyton-Brown, K., 2008. *Multiagent Systems: Algorithmic, Game-Theoretic, and Logical Foundations*. Cambridge University Press.

Simmons, R.E., Kolberg, H., Braby, R., Erni, B., 2015. Declines in migrant shorebird populations from a winter-quarter perspective. *Conservation Biology* **29**, 877–87.

Simon, H.A., 1979. Rational decision making in business organizations. *American Economic Review* **69**, 493–513.

Skonhoft, A., Vestergaard, N., Quaas, M., 2012. Optimal harvest in an age structured model with different fishing selectivity. *Environmental and Resource Economics* **51**, 525–44.

Smith, R.B.W., Shogren, J.F., 2001. Protecting species on private lands. In Shogren, J., Tschirhart, J. (eds.), *Protecting Endangered Species in the United States*. Cambridge University Press, pp. 326–42.

Smith, R.B.W., Shogren, J.F., 2002. Voluntary incentive design for endangered species protection. *Journal of Environmental Economics and Management* **43**, 169–87.

Soltani, A., Sankhayan, P.L., Hofstad, O., 2014. A dynamic bio-economic model for community management of goat and oak forests in Zagros, Iran. *Ecological Economics* **106**, 174–85.

Spencer, R.-J., van Dyke, J.U., Thompson, M.B., 2017. Critically evaluating best management practices for preventing freshwater turtle extinctions. *Conservation Biology* **31**, 1340–9.

Starfield, A.M., 1997. A pragmatic approach to modeling for wildlife management. *Journal of Wildlife Management* **61**, 261–70.

Starfield, A.M., Smith, K.A., Bleloch, A.L., 1990. *How to Model It: Problem Solving for the Computer Age*. McGraw-Hill.

Stauffer, D., 2007. Social applications of two-dimensional Ising models. arXiv:0706.3983v1 [physics.soc-ph], 27 June 2007.

Stauffer, D., Aharony, A., 1994. *Introduction to Percolation Theory*. Taylor & Francis.

Stern, N.H., 2007. *The Economics of Climate Change: The Stern Review*. Cambridge University Press.

Stillman, R.A., Railsback, S.F., Giske, J., Berger, U., Grimm, V., 2015. Making predictions in a changing world: The benefits of individual-based ecology. *BioScience* **65**, 140–50.

Stintzing, S., Zwerger, W., 1997. Ginzburg–Landau theory of superconductors with short coherence length. *Physical Review B* **56**, 9004.

Stoeven, M.T., 2014. Enjoying catch and fishing effort: The effort effect in recreational fisheries. *Environmental and Resource Economics* **57**, 393–404.

Strange, N., Thorsen, B. J., Bladt, J., 2006. Optimal reserve selection in a dynamic world. *Biological Conservation* **131**, 33–41.

Sturm, A., Drechsler, M., Johst, K., Mewes, M., Wätzold, F., 2018. DSS-Ecopay – A decision support software for designing ecologically effective and cost-effective agri-environment schemes to conserve endangered grassland biodiversity. *Agricultural Systems* **161**, 113–16.

Sukhdev, P., Wittmer, H., Miller, D., 2014. The Economics of Ecosystems and Biodiversity (TEEB): Challenges and responses. In Helm, D., Hepburn, C. (eds.), *Nature in the Balance: The Economics of Biodiversity*. Oxford University Press.

Surun, C., Drechsler, M., 2018. Effectiveness of tradable permits for the conservation of metacommunities with two competing species. *Ecological Economics* **147**, 189–96.

Tadelis, S., 2013. *Game Theory: An Introduction*. Princeton University Press.

Tahvonen, O., Quaas, M.F., Schmidt, J.O., Voss, R., 2013. Optimal harvesting of an age-structured schooling fishery. *Environmental and Resource Economics* **54**, 21–39.

Taubert, F., Fischer, R., Groeneveld J., et al., 2018. Global patterns of tropical forest fragmentation. *Nature* **554**, 519–22.

TEEB 2010a. *The Economics of Ecosystems and Biodiversity: Mainstreaming the Economics of Nature: A Synthesis of the Approach, Conclusions and Recommendations of TEEB.* www.teebweb.org/our-publications/teeb-study-reports/synthesis-report/ #.Ujr2cX9mOG8 (last accessed 10 October 2019).

TEEB 2010b. *The Economics of Ecosystems and Biodiversity: Ecological and Economic Foundations.* Edited by P. Kumar. Earthscan. www.teebweb.org/our-publica tions/teeb-study-reports/ecological-and-economic-foundations/#.Ujr1xH9mOG8 (last accessed 10 October 2019).

Teel, T.L., Anderson, C.B., Burgman, M.A., et al., 2017. Publishing social science research in *Conservation Biology* to move beyond biology. *Conservation Biology* **32**, 6–8.

Tenan, S., Iemma, A., Bragalanti, N., et al., 2016. Evaluating mortality rates with a novel integrated framework for nonmonogamous species. *Conservation Biology* **30**, 1307–19.

Theesfeld, I., MacKinnon, A., 2014. Giving birds a starting date: The curious social solution to a water resource issue in the U.S. West. *Ecological Economics* **97**, 110–19.

Thierry, H., Sheeren, D., Marilleau, N., et al., 2015. From the Lotka–Volterra model to a spatialised population-driven individual-based model. *Ecological Modelling* **306**, 287–93.

Thomas, B.A., Azevedo, I.L., 2013. Estimating direct and indirect rebound effects for U.S. households with input-output analysis, Part 1: Theoretical framework. *Ecological Economics* **86**, 199–210.

Thompson, H., 2013. Resource rights and markets in a general equilibrium model of production. *Environmental and Resource Economics* **56**, 131–9.

Thulke, H.-H., Grimm, V., Müller, M.S., et al., 1999. From pattern to practice: A scaling-down strategy for spatially explicit modelling illustrated by the spread and control of rabies. *Ecological Modelling* **117**, 179–202.

Tietenberg, T.H., 2006. *Emissions Trading: Principles and Practice.* Resources for the Future, Washington, DC.

Tietenberg, T., Lewis, L., 2018. *Environmental and Natural Resource Economics.* Routledge, 11th ed.

Tilman, D., 1994. Competition and biodiversity in spatially structured habitats. *Ecology* **75**, 2–16.

Tilman, D., Kareiva, P. (eds.), 1997. *Spatial Ecology: The Role of Space in Population Dynamics and Interspecific Interactions.* Princeton University Press.

Todd, C.R., Lintermans, M., 2015. Who do you move? A stochastic population model to guide translocation strategies for an endangered freshwater fish in south-eastern Australia. *Ecological Modelling* **311**, 63–72.

Touza, J., Drechsler, M., Smart, J.C.R., Termansen, M., 2013. Emergence of cooperative behaviours in the management of mobile ecological resources. *Environmental Modelling & Software* **45**, 52–63.

Tuljarpurkar, S., Caswell, H., 1997. *Structured-Population Models in Marine, Terrestrial, and Freshwater Systems.* Springer.

Turner, B.L., Wuellner, M., Nichols, T., et al.., 2016. Development and evaluation of a system dynamics model for investigating agriculturally driven land transformation in the North Central United States. *Natural Resource Modeling* **29**, 179–228.

Uehara, T., 2013. Ecological threshold and ecological economic threshold: Implications from an ecological economic model with adaptation. *Ecological Economics* **93**, 374–84.

Uehara, T., Nagase, Y., Wakeland, W., 2016. Integrating economics and system dynamics approaches for modelling an ecological-economic system. *Systems Research and Behavioral Science* **33**, 515–31.

Ulbrich, K., Drechsler, M., Wätzold, F., Johst, K., Settele, J., 2008. A software tool for designing cost-effective compensation payments for conservation measures. *Environmental Modelling & Software* **23**, 122–3.

van Kooten, G.C., Johnston, C.M.T., 2016. The economics of forest carbon offsets. *Annual Review of Resource Economics* **8**, 227–46.

van Teeffelen, A., Opdam, P., Wätzold, F., et al., 2014. Ecological and economic conditions and associated institutional challenges for conservation banking in dynamic landscapes. *Landscape and Urban Planning* **130**, 64–72.

van Winkle, C., Jager, H.I., Railsback, S.F., et al., 1998. Individual-based model of sympatric populations of brown and rainbow trout for instream flow assessment: Model description and calibration. *Ecological Modelling* **110**, 175–207.

Varian, H.R., 2010. *Intermediate Microeconomics: A Modern Approach*. Norton & Company, 8th ed.

Vellend, M., 2016. *The Theory of Ecological Communities*. Princeton University Press.

Verhagen, W., van der Zanden, E.H., Strauch, M., van Teeffelen, A.J.A., Verburg, P.H., 2018. Optimizing the allocation of agri-environment measures to navigate the trade-offs between ecosystem services, biodiversity and agricultural production. *Environmental Science & Policy* **84**, 186–96.

Verhoef, H.A., Morin, P.J., 2009. *Community Ecology: Processes, Models, and Applications*. Oxford University Press.

Vincent, T., 2005. *Evolutionary Game Theory, Natural Selection, and Darwinian Dynamics*. Cambridge University Press.

Visconti, P., Joppa, L., 2014. Building robust conservation plans. *Conservation Biology* **29**, 503–12.

VNP 2018. Bayerisches Vertragsnaturschutzprogramm, Verpflichtungszeitraum 2018–2022: Maßnahmenübersicht. www.stmelf.bayern.de/mam/cms01/agrar politik/dateien/massnahmenuebersicht_vnp.pdf (last accessed 10 October 2019).

Voinov, A., 2008. *Systems Science and Modeling for Ecological Economics*. Academic Press.

Volterra, V., 1931. Variations and fluctuations of the number of individuals in animal species living together. In Chapman, R.N. (ed.), *Animal Ecology with Special Reference to Insects*. McGraw-Hill, pp. 409–48.

von Neumann, J., Morgenstern, O., 2007. *Theory of Games and Economic Behavior: 60th Anniversary Commemorative Edition*. Princeton University Press.

de Vries, F.P., Hanley, N., 2016. Incentive-based policy design for pollution control and biodiversity conservation: A review. *Environmental and Resource Economics* **63**, 687–702.

Wakano, J.Y., Nowak, M.A., Hauert, C., 2009. Spatial dynamics of ecological public goods. *Proceedings of the National Academy of Sciences of the USA* **106**, 7910–14.

Walker, B., Holling, C.S., Carpenter, S.R., Kinzig, A., 2004. Resilience, adaptability and transformability in social-ecological systems. *Ecology and Society* **9**(2), 5.

Walker, S., Brower, A.L., Stephens, R.T.T., Lee, W.G., 2009. Why bartering biodiversity fails. *Conservation Letters* **2**, 149–57.

Wang, Y., Atallah, S., Shao, G., 2017. Spatially explicit return on investment to private forest conservation for water purification in Indiana, USA. *Ecosystem Services* **26**, 45–57.

Warwick-Evans, V., Atkinson, P.W., Walkington, I., Green, J.A., 2018. Predicting the impacts of wind farms on seabirds: An individual-based model. *Journal of Applied Ecology* **55**, 503–15.

Watmough, J., Edelstein-Keshet, L., 1995. Modelling the formation of trail networks by foraging ants. *Journal of Theoretical Biology* **176**, 257–371.

Wätzold, F., Drechsler, M., 2014. Agglomeration payment, agglomeration bonus or homogeneous payment? *Resource and Energy Economics* **37**, 85–101.

Wätzold, F., Drechsler, M., Armstrong, C.W., et al., 2006. Ecological-economic modeling for biodiversity management: Potential, pitfalls, and prospects. *Conservation Biology* **20**, 1034–41.

Wätzold, F., Drechsler, M., Johst, K., Mewes, M., Sturm, A., 2016. A novel, spatiotemporally explicit ecological-economic modeling procedure for the design of cost-effective agri-environment schemes to conserve biodiversity. *American Journal of Agricultural Economics* **98**, 489–512.

Wätzold, F., Lienhoop, N., Drechsler, M., Settele, J., 2008. Estimating optimal conservation in the context of agri-environmental schemes. *Ecological Economics* **68**, 295–305.

Wätzold F., Schwerdtner, K., 2005. Why be wasteful when preserving a valuable resource? A review article on the cost-effectiveness of European biodiversity conservation policy. *Biological Conservation* **123**, 327–38.

Weibull, J.W., 1995. *Evolutionary Game Theory*. MIT Press.

Wende, W., Albrecht, J., Darbi, M., et al., 2018. Germany. In Wende, W., Tucker, G.-M., Quétier, F., Rayment, M., Darbi, M., (eds.), *Biodiversity Offsets: European Perspectives on No Net Loss of Biodiversity and Ecosystem Services*. Springer, pp. 123–56.

White, B., Hanley, N., 2016. Should we pay for ecosystem service outputs, inputs or both? *Environmental and Resource Economics* **63**, 765–87.

Wiegand, T., May, F., Kazmierczak, M., Huth A., 2017. What drives the spatial distribution and dynamics of local species richness in tropical forest? *Proceedings of the Royal Society B* **284**, 20171503.

Wiki, 2018. *List of Games in Game Theory*. en.wikipedia.org/wiki/List_of_games_in_game_theory (last accessed 10 October 2019).

Wilson, J.D., Hopkins, W.A., 2013. Evaluating the effects of anthropogenic stressors on source-sink dynamics in pond-breeding amphibians. *Conservation Biology* **27**, 595–604.

Wissel, C., Zaschke, S.-H., 1994. Stochastic birth and death processes describing minimum viable populations. *Ecological Modelling* **75**/76, 193–201.

Wissel, S., Wätzold, F., 2010. A conceptual analysis of the application of tradable permits to biodiversity conservation. *Conservation Biology* **24**, 404–11.

World Commision on Environment and Development (WCED), 1987. *Our Common Future*. Oxford University Press.

Wu, J.J., Boggess, W.G., 1999. The optimal allocation of conservation funds. *Journal of Environmental Economics and Management* **38**, 302–21.

Zeigler, S.L., Walters, J.R., 2014. Population models for social species: Lessons learned from models of red-cockaded woodpeckers (*Picoides borealis*). *Ecological Applications*, **24**, 2144–54.

Zhu, B., Wang, K., Chevallier, J., Wang, P., Wei, Y.-M., 2015. Can China achieve its carbon intensity target by 2020 while sustaining economic growth? *Ecological Economics* **119**, 209–16.

Zinck, R.D., Grimm, V., 2009. Unifying wildfire models from ecology and statistical physics. *American Naturalist* **174**, E170–84.

Zinck, R.D., Johst, K., Grimm, V., 2010. Wildfire, landscape diversity and the Drossel–Schwabel model. *Ecological Modelling* **221**, 98–105.

Index